AF788803

JAN 2014

Meeting the Challenge: The Hexagon KH-9 Reconnaissance Satellite

Meeting the Challenge: The Hexagon KH-9 Reconnaissance Satellite

Phil Pressel

LIBRARY OF FLIGHT

Ned Allen, Editor-in-Chief
Lockheed Martin Corporation
Bethesda, Maryland

Published by
American Institute of Aeronautics and Astronautics, Inc.
1801 Alexander Bell Drive, Reston, VA, 20191-4344

Front cover photograph of the Hexagon as well as Figures 2.5, 2.6, 8.1, 8.4, 8.5, and 8.6 are (c) Jon Aspinwall. Reprinted with permission.
Chapter 18 photographs reprinted with permission of Lorenzo Hagerty.

American Institute of Aeronautics and Astronautics, Inc., Reston, Virginia

1 2 3 4 5

Library of Congress Cataloging-in-Publication Data
Record on file.

Copyright © 2013 by the American Institute of Aeronautics and Astronautics, Inc. All rights reserved. Printed in the United States of America. No part of this publication may be reproduced, distributed, or transmitted, in any form or by any means, or stored in a database or retrieval system, without the prior written permission of the publisher.

Data and information appearing in this book are for informational purposes only. AIAA is not responsible for any injury or damage resulting from use or reliance, nor does AIAA warrant that use or reliance will be free from privately owned rights.

ISBN 978-1-62410-203-5

Dedication

I wish to dedicate this book to all of the unsung heroes of this fantastic achievement, my colleagues at Perkin-Elmer and to the company itself. There were hundreds of employees who worked on the project. I praise their outstanding skills, inventiveness, smarts, and dedication. This includes managers, engineers (electrical, mechanical, systems, optical, test, quality, manufacturing, and reliability), technicians, draftsmen, contracts, purchasing, secretaries, administrators, personnel program office, testing, optical metrology, machine shops, shipping and handling, field operations, and support personnel from all departments.

Also to be praised for their engineering, scientific smarts, and dedication are the people from our customer; the CIA, the NRO, and the Air Force, as well as to associated contractors and advisors to the agencies.

It is unfortunate that it took many years for the program to become declassified and its description revealed. Many participants are no longer alive and it is probable that their families and friends will never know the crucial role that each played in one of the most important achievements for U.S. national security. While I was working on this program, I believed that most of us thought that it was important and a privilege to work on it. Now I also believe that it took a special ingenuity and dedication by inimitable people to have accomplished this magnificent achievement.

Phil Pressel

A Note from the Editor-in-Chief

Perhaps as significant as the discovery and development of the laws of science and engineering that enabled the aircraft and spacecraft of the 20th century, was the simultaneous discovery of the geopolitical stabilizing influence of application of those technological marvels to Intelligence, Surveillance, and Reconnaissance (ISR). The crowning jewel of ISR in the 20th century was surely the "Keyhole" series of overhead assets developed and operated by the U.S. Intelligence Community. In this remarkable recounting of the Hexagon project, perhaps the most stunning technological tour de force of the KH series, Phil Pressel, a principal engineer on that project, draws together the personal and engineering stories behind the history. Every U.S. President from Eisenhower to Reagan, who famously commented "trust but verify," depended on these overhead ISR assets to "verify" treaty compliance and to independently assess the truth behind the many threats of Cold War geopolitics. So this well-written story, one of the most significant of the 20th century and until recently among the deepest secrets (declassified only in 2011), belongs on the Library of Flight bookshelf.

The Library of Flight is part of the comprehensive portfolio of information services from the American Institute of Aeronautics and Astronautics. It extends the Institute's offerings with the best material in a growing variety of topics in aerospace from policy, to histories, to law, management, and beyond. The Library of Flight documents the crucial role of aerospace in enabling, facilitating, and accelerating our technological society and its global commerce, communication, and defense.

Ned Allen
Bethesda, Maryland
July 2013

Contents

Preface

The following are four of the most memorable dates over a 20-year period of a most exciting, successful, and challenging engineering exploit known as the Hexagon, a reconnaissance satellite.

The first was in May 1966 when Chester Nimitz, Jr., CEO of the Perkin-Elmer Corporation and the son of the famous World War II admiral, stood up at the end of a final proposal presentation to our customer, the CIA, and put his foot up on a table and said, "We want this f———g job and we're gonna get every f———g agency and every f———g engineer from here to Florida. We recognize the importance to national security and we're capable of doing the job."

The second date was October 10, 1966 when Mike Maguire, our program manager, and Dick Werner, our senior vice president, walked into our work area where gathered the 30 or so of us who had worked so hard on the proposal. As Dick took out a cigar from inside his jacket, he said "We won."

The third date was June 15, 1971. Sitting next to Mike Maguire with several others in our war room, I listened on our closed circuit audio system to the launch controller countdown. Our first launch—a Titan 3D rocket lifted off with about 3 million pounds of thrust carrying an approximate 30,000 pounds of payload, 60 feet long, and 10 feet in diameter. A period of silence followed, then periodic updates as to altitude and speed and eventually confirming that the Hexagon film reconnaissance system was in orbit.

The last date was April 18, 1986. By now, the countdown was familiar. But this time a noisy and powerful exhaust came from the rocket as it launched off the pad at Vandenberg Air Force Base in California. Then disaster happened. Before the rocked reached 1000 feet, there was a fiery explosion.

What was Hexagon?

It was the last film-based orbiting photo-reconnaissance satellite. It provided invaluable intelligence information for the government of the United States. The film capsules were periodically returned to Earth in reentry vehicles that were caught over the Pacific Ocean by Air Force C-130 planes. Eventually, newer reconnaissance satellites used solid-state technology that send the photo data back to earth electronically in real time. Our customer was the CIA and all of the work was top secret.

My Story

I have been asked recently by a number of people: how did a World War II Holocaust survivor become involved with designing spy cameras for the CIA? As they say in the real estate business, the answer is "location, location, location." In my case, it was location (being at the Perkin-Elmer Corporation) and timing (being there at the start of the program). It also involved good luck. I had the right experience and was able to obtain my security clearance easily. As the work progressed I developed a great amount of respect for this important program. It instilled in me a great amount of pride.

I was lucky to have been admitted to one of three specialty high schools in New York City, Stuyvesant High School, where I was exposed to science, math, and technology. After graduating from New York University as a mechanical engineer, my first job was at Radio Corporation of America in Camden, New Jersey. Inasmuch as it dealt mainly with electrical devices it was not a great place to get good mechanical engineering experience. I eventually got a good job at the Avco Corporation in Wilmington, Massachusetts doing design and analysis of reentry vehicles.

I had always enjoyed doing mechanical design and specifically stress analysis, at which I thought I was good. The men that I worked with at Avco were the true experts. It was there that I really learned how to do stress analysis—determining the stress and deflections of structures when undergoing g loads, vibration, and changes in thermal environment. This learning was to serve me well for the rest of my whole career. After three years of working there, business tapered off and I could see layoffs coming.

One of the companies that I interviewed with was Perkin-Elmer in Connecticut. My interviewer was Mike Krim. He was a charismatic, very nice, smart, and slightly "odd professorial" fellow. He explained that the company designed and fabricated optical instruments, such as large airborne cameras, telescopes, and other sophisticated instruments. He showed me around and made me feel quite welcome. I became hopeful of receiving an offer. I did receive one and accepted and planned to start work at the beginning of May 1965.

I was told that eventually I was slated to work on a classified program and that I would have to fill out a personnel security questionnaire (PSQ). It was many pages long and asked for my life's history. This included all addresses for wherever I had lived, all jobs I ever held, and all the places that I had traveled to. Since I was an immigrant from Belgium and France, and had traveled a lot even after arriving in the United States in 1946, I had a lot of information to gather. It was a real pain, but I did it. After only 4 months I received my top secret clearance. Periodically for the rest of my career I needed to update my PSQ to maintain my clearance. I found out that I would be working on the design of a large high-resolution camera that was to orbit the earth and spy on our enemies. My initial reaction was, "Wow, this sounded like important and interesting work." Later I found out that the customer was the CIA.

My first real assignment was to create a mechanical design for the optical bar, the actual optical assembly or camera, for the Hexagon system. It was only a study at the time. A configuration for the location of the mirrors and lenses had been formulated by the optical designer and given to me. I was to analyze the optics and structure for stresses and deflections generated during launch and due to changes in thermal environment, and then start to design supporting structure for the optics. This became quite a complicated job and became my primary responsibility later when we actually won the job. I was able to use much of my mechanical engineering education and do the things that I enjoyed the mostdesign and stress analysis. I also learned a lot of new things, especially concerning optics. For one thing, I learned to deal in very small deflections for the optics. I had been used to calculating deflections in fractions of inches and tilts in degrees. Now I was dealing with microns (a millionth of a meter) and angles of arc seconds.

Success

I remember October 10, 1966 very well. This was approximately 4 months after the proposal was submitted to our customer. All of us were told to gather at 10 a.m. in the large engineering room. This was across the street from one of the two main plants of the company. We were located there because it was an isolated and secure area.

Shortly after 10, in walked Dick Werner (our group's vice president), Mike Maguire (our program leader), and Charley Hall (who handled contracts). They were each dressed in stylish suits. In those days all of us wore ties and jackets, although the latter were soon discarded as each day progressed. As they reached the front of the room, Dick reached into his right inside jacket pocket and took out one of the longest cigars that I had ever seen. I immediately concluded that we had won the job. The first words out of his mouth were "We won." A great cheer went up from all of us. Dick and Mike then each spoke a few words of praise for the great team effort and wished us success in this new adventure.

As soon as the meeting broke up, some of us immediately went to our desks and picked up the telephone. Many, of course, called their wives. We had no female engineers or scientists in those days to call their husbands. Some were added much later. Some employees called their stockbrokers to give orders to buy stock in our company for whatever amounts of money each could afford. I was one of those who purchased a very small number of shares, as I could not afford much.

Of course this was illegal as it was trading using insider information. Later in the day and the next day a secretary was told to go to everyone and ask if they had purchased shares and how many. This list was eventually given to our legal department. All those involved expected to be reprimanded and possibly made to sell or perform some other financial transaction to void the purchases. No word was ever communicated to us and nothing was ever done to anyone. I assume the total number of stocks purchased was not a significant amount. It turned out to be a lucrative investment: the company stock split seven times in the next dozen years.

Why did I Write This Book?

I wrote the book as a labor of love to honor the people who worked on the program and the organizations that participated, especially my company, the Perkin-Elmer Corporation (now United Technologies). It was a career job for me and I am sure it was for many other participants. It was an incredible teamwork effort that resulted in success that exceeded the requirements, thus "meeting the challenge."

The main purpose of this book is to reveal, for the first time, details of the intricacies of the program, its development, and production. The book also contains some human-interest stories about its participants and their feelings and experiences. Their own words reveal how awesome the challenges were. Other publications contain partial information about Hexagon but none are written by anyone who actually worked on it. *Meeting the Challenge: The Hexagon Reconnaissance Satellite* is truly an authoritative document.

How did I Write This Book?

Prior to my retirement in 1999 I had planned to write a book about how I and my parents survived the Holocaust. We were living in Belgium when the German army invaded in May 1940. We, being Jewish, were then forced to be on the run from the Nazis for the next 5 years in France. I wrote *They Are Still Alive* so that my children and grandchildren would learn about their heritage. The book was published in 2004 by Dorrance Publications.

While writing the memoir, I discovered that I enjoyed the writing process. What else could I write that I knew a lot about? The Hexagon program—I worked on the job from its beginning to its end.

I first had to write to the CIA to obtain permission because the program was still highly classified. In March 2004 I wrote to them asking if I could write about Hexagon and gave them a general outline of the proposed document. They called me approximately a month later and invited me to come and speak directly with Paul Noel Chretien, the chairman of the CIA Publication Review Board.

In June 2004, I visited with Paul in his office. I told him my plans were to explain in detail how the whole camera and its individual major components worked, including drawings. I also told him that some of my colleagues on the program would be writing portions that were their expertise, as well as to have all of us write personal stories about events. He explained the approval process that I would have to go through and okayed my approach. I gave him the names of my colleagues and he subsequently wrote a letter to each of us giving permission to write as long as the writings were kept secure and that no material could be seen by noncleared people or published until the CIA reviewed the document and the program itself was declassified.

I decided that all of the writings be sent to me. In order to keep them secure, I took several steps. First, I had a safe bolted to the floor of my bedroom closet in which I kept all documents. Most importantly I obtained a laptop computer that was totally and exclusively dedicated to writing this book, and I added an encryption program. I never connected this laptop to the Internet, so hacking was never an issue. I wrote the book on an external hard drive and backed it up on a thumb drive, both of which I kept in the safe when not working on them. Eventually a security agent from the CIA came to my home to inspect my security arrangements and found them to be satisfactory.

About Hexagon

The program was called the KH-9 orbital camera system and known by its code name Hexagon. It was known in the media as Big Bird. Technical details in this book came from the engineers and scientists responsible for major subassemblies or technical disciplines. All of the information is derived from our collective memories because detailed documentation in the form of specifications, drawings, report, and analyses were unavailable during the writing of this book due to the fact that they were still classified.

The program began with a small group of us who worked on a study for about a year before we were asked by the CIA to write the proposal, which was an extremely intense 6-week process. The program contract was awarded to us at the Perkin-Elmer Corporation. It was to become the most successful spy satellite the United States had ever placed in orbit around Earth. At that time, the entire Perkin-Elmer Corporation was doing about $50 million dollars per year of total revenues. The contract would eventually be extended to over 20 years and grow enough financially to help place Perkin-Elmer on the Fortune 500 list.

The last scheduled launch of a Hexagon system was April 18, 1986. The rocket exploded and that was the end of the fantastic and hugely successful Hexagon program. It was the only one out of 20 systems not to fly. That day I was walking in the main hall of our building when one of the senior managers told me what had just happened. I rushed to my office in shock, and out came the tears. I had been intimately involved with the program since its beginning over 20 years earlier; it was my baby.

This reconnaissance satellite was probably the most complicated system ever put into space and was an engineering marvel. Before declassification, no details of its design aspects and features had been disclosed other than speculation of its accuracy. In 1972, President Nixon signed the first SALT agreement based on Big Bird as the technical means of verifying or monitoring compliance.

In October 1978, President Carter in a public address to honor several astronauts at the Kennedy Space Center acknowledged the existence and importance of reconnaissance satellites and praised them as an immensely valuable asset for the security of the United States: "Photo-reconnaissance satellites have become an important stabilizing factor in world affairs for the monitoring of arms control agreements. They make an immense contribution to the security of all nations. We should continue to develop them." Those of us working on the project at the time were called into group meetings and told that the speech would be mentioned in the news later that day and that we were not to acknowledge or confirm President Carter's statement, or even the existence of the program, as it was highly classified.

Besides being incredibly complicated, the program resulted in many new technical innovations that could be called "world records" if they were considered like an Olympic sport. Among others these included:

- The ability to control film traveling at very high speed (200 inches per second)
- A huge quantity of film (60 miles) used for each mission
- The world's largest spherical thermal vacuum chamber used to test the system
- Stereo photography of the Earth's entire surface

- The development and use of new and sophisticated electronics such as light-emitting diodes (LEDs) and brushless motors

No one was ever allowed to talk about it with family or friends or with anyone who did not have the proper security clearance. We could not "boast or brag," share our pride, or discuss our accomplishments and successes for the work we did. Secrecy became second nature to us, and we got proficient talking among ourselves in technical abbreviation codes relevant to the program.

Twenty-five years after the program ended, it was finally declassified on September 17, 2011. So finally we could tell our families what we did. I and my wife were privileged to have been invited to the declassification ceremony in Washington, D.C. Dozens of my former colleagues also attended along with representatives from other major contractors (Lockheed-Martin, McDonnell Douglas, Eastman Kodak, Itek, and General Electric) plus representatives from the government including the military.

The biggest thrill for me was to see my old colleagues and the exhibits of the actual Gambit (another earlier reconnaissance satellite) and Hexagon systems. The Hexagon system on display was the development model, a 60-feet long vehicle containing the compartment for the film supply, the whole two-camera assembly, and several of the reentry vehicles. The D-model had been used throughout the program as a fully working system to simulate problems that developed on the ground and in orbit on real flight units and was used to help resolve the problems. It was so much fun to discuss the hardware with my colleagues. The model is now at the Wright-Patterson Air Force Museum in Dayton, Ohio.

Another celebration was held on October 24, 2011 at Goodrich Corporation (formerly Perkin-Elmer, now known as United Technologies) where we all worked in Danbury, Connecticut. The company was gracious to invite all of the Hexagon alumni that they could reach. It was organized by Charlie Cox, Special Projects Director with assistance from Vic Abraham, retired Hexagon Program Director and Becky Zepf, of Goodrich Human Relations.

I have put the very detailed technical engineering explanations for some chapters in associated appendixes for the sake of simplification. These details would be of interest to the engineers and specialists among the public who care to delve more deeply into the subject.

Throughout the book I have interspersed some sections and chapters that are personal stories of my colleagues' beginnings on the program, their own interpretation of how the system worked, and discussions of some of their experiences. There is some repetition in these chapters, but as it was an extremely complicated system some repetition will make it easier to understand.

Much of the information in the book can be verified in several National Reconnaissance Office (NRO) documents that can be found on NRO.gov under Hexagon September 17, 2011. This book was reviewed by the CIA/NRO and was approved for publication.

Phil Pressel
San Diego, California
August 2013

Acknowledgements

The decision to write this book was made in 2004—years before the Hexagon program was declassified. Permission had to be obtained from the CIA to write it. Before the book could be published, instructions were that the manuscript had to be approved by the CIA and that some parts of the book might be stricken. Any errors in the book are mine.

There are many people whom I want to thank for contributing and helping make this book a reality. I first want to thank Paul Noel Chretien, former Chairman of the CIA's Publications Review Board. I spent many hours with him on the phone and in his office in Washington, D.C. receiving explanation of the declassification process and excellent advice on how to proceed with the book and what steps it would take to get the Hexagon program declassified. Paul guided me throughout the process until he left the board. Also helping me was his assistant and deputy chairman of the review board, Richard Florence.

Once the Hexagon program was declassified on September 17, 2011, I was helped greatly by Richard Puhl, Chairman of the CIA Publication Review Board; his assistant David Vintrello; and Steven Glenn, Chief of Information Access and Release Team of the NRO.

It was difficult for all of my colleagues at Perkin-Elmer to sit down and start writing about things that happened so many years ago. There was no technical record available to us due to the classified nature of the job. In most cases it was approximately 40 years ago when many of us began work on Hexagon. Trying to get started to fill up an empty page was very hard. The more that we talked and wrote, however, the more our memory recalled details stored way back in our minds. It is amazing how some mental stimulation brings things back. One of the joys of writing this book was to relive the job of a lifetime and reestablish contact with former colleagues. They and everyone who worked on it formed an outstanding team made up of incredibly talented people. I salute them all.

I specifically want to thank and acknowledge the Perkin-Elmer people who helped to write portions of the book. Without their intimate knowledge of their specialties, be it a technical discipline, or describing the design of an actual major piece of hardware and processes of integration from the start to finish, the book would have been impossible to complete. Some of the portions are quite large and those colleagues received a chapter credit. I specifically want to cite Dick Carricato and Sam Palasciano for their advice and editing skills.

Others who contributed in the form of taped interviews; by supplying information or photographs; or gave me their insight and reviewed parts of the manuscript include Jon Aspinwall, Peter Green, Len Farkas, Bob Jones, Stan Karachuk, Paul Convertito, Bob Perliss, Ed Gaal, Bob Landsman, Frank Grabowski, Mike Maguire, Mike Mazaika, Linc Endelman, George Manolis, Bob Meyers, Herb Yanowitz, Joe Vollaro, Frank Sileo, Kent Meserve, Rich Molaskey, Dick Paltauf, Paul Yoder, Bill Lynch, Bud Rigby, Dick Babish, and Jim Miller.

I also want to acknowledge in general the contributions of many talented people from the CIA staff that was part of the Directorate of Science and Technology. Also acknowledged are the employees of the TRW Corporation and the Aerospace Corporation who provided guidance and consulting services to the CIA, and to Lockheed (now Lockheed-Martin), the overall integrator of all flight systems. The same technical skills, dedication, and smarts that were exhibited by the Perkin-Elmer staff apply to these organizations. Many were outstanding scientists, engineers, and managers who displayed sound thinking and were a good sounding board for ideas. They were especially helpful in the early days during the development of the program. Kudos also goes to many vendors and subcontractors who designed and fabricated critical equipment ranging from electronics to reentry vehicles.

Similar words of praise apply to the staff of the organization that eventually took over the program once it was operational in the 1972 time frame, namely the Air Force. Their talented people took over the Herculean task of running an already established program. They facilitated the logistics, were involved in some improvements, and ran all of the launches. I also want to thank Lorenzo Hagerty, U.S. Navy Executive Officer on the Apache and the Navigator for the entire unit (Whitesands, Trieste, and Apache) during the recovery of a reentry vehicle sunk in the Pacific and retrieved by the Trieste. I thank him for the photographs he took during this event.

I want to thank my editor Dwayne Day, whose expertise and interest in space technology as well as his experience in writing and editing made this book less wordy and better organized. Dwayne Day works for the National Research Council where he has directed studies for NASA and the U.S. Air Force on NASA's strategic direction, the future of planetary exploration, the threat to Earth from asteroids, the role of the astronaut corps, and satellite tracking. He previously worked for The Space Policy Institute and the

Congressional Budget Office as well as serving as an investigator on the Columbia Accident Investigation Board.

Lastly I want to thank my family. I thank my wife Pat for all of her support over the 9 years that I have been working on this book. She put up with my needs, including helping me with computer issues and how to structure the book's contents (without knowing anything about the program) as well as giving me the time and space to do my work. I cherish and love her, and I know that I could not have written this book without her. I am eternally grateful for her patience, caring, support, and devotion during all our years together.

My wife's nephew and his wife, Michael and Nancy Connors, deserve special credit. In 2006, only 2 years into writing this book, Nancy's gift to me of one of her kidneys, along with her and her husband's support, afforded me the good health that enabled me to continue writing. They gave me the gift of life and I am forever indebted to them.

The word teamwork is bandied about a great deal. In the case of Hexagon, perhaps the most complicated satellite ever put up in orbit, it was truly a team effort by incredibly talented people in all phases of expertise. They worked together extremely well and without their skills and dedication it would not have happened. Their pride is well deserved.

I also want to acknowledge people involved from the following organizations that were important contributors to the program and had close relationships with us at Perkin-Elmer.

CIA

Roy Burks
John Crowley
Bob Kohler
John McMahon
Don Patterson

Air Force

Colonel Les McChristian, Program Director
Colonel Larry Cress, Program Director
Colonel Tom O'Neill, onsite at Perkin-Elmer as Air Force representative
Colonel Dave Berganini
Colonel Gayney Best
Colonel Clark Carlin
Colonel Ray Anderson
Colonel Dave Raspet

Nro

Frank Buzard, Director of Program A
Steve Glenn, Chief Information Access and Release Team

Industry

Edwin Land, scientist, developer of the Polaroid Land Camera, and member of scientific advisory groups to several U.S. presidents

Perkin-Elmer

Rod Scott, Chief Scientist and photo-reconnaissance expert
Bob Sorensen, President (during most of the program)
Chester Nimitz, Jr., CEO (during most of the program)
Don Cowles, inventor of the twister
Charlie Karatzas, Manager of Mechanical Engineering, then Director of Engineering
Len Molaskey, Manager of Mechanical Engineering and designer of reentry retrieval hardware
Rich Molaskey, Administrator of Document Control
Arnie Wallace, all-around engineer and devil's advocate

Others

Scott Campion, computer expert who set up an encryption program for my manuscript
Chris Hatch, design engineer formerly at the Quartus Engineering Company who helped me with CAD
Ken Paulson, IT expert at Quartus Engineering who helped me with computer issues

Abbreviations

Many of the abbreviations listed below are spelled out in the book. They are included here to demonstrate the many abbreviated terms used in conversation and documentation to maintain the security of the Hexagon program.

AFB	Air Force Base
ASD	Aerospace System Division
BFL	back focal length
BR	builder roller
CAD	computer-aided design
CCD	charged coupled device
CDR	critical design review
CEO	Chief Executive Officer
CIA	Central Intelligence Agency
CTS	coarse tension sensor
DC	drive capstan (or direct current)
DDAS	digital data acquisition system
D/O	deorbit
DOD	Department of Defense
DS&T	Directorate of Science and Technology
ECR	engineering change request
EK	Eastman Kodak
EMI	electromechanical interference
EOD	Electro-Optical Division
ESD	emergency shutdown
FBI	Federal Bureau of Investigation
FD	film drive
FEA	finite element analysis
FEV	film exit vestibule
FFP	fine film path
FMECA	failure mode electrical criticality analysis
FP	film path or focal plane

FPS	film path simulator
GE	General Electric
GN2	gaseous nitrogen
IC	integrated circuit
ICBM	intercontinental ballistic missile
IESD	independent emergency shutdown
IFTS	input fine tension sensor
IMC	image motion compensation
LED	light-emitting diode
LEM	lunar excursion module
LEO	low Earth orbit
LMSC	Lockheed Missile and Space Corporation
LO	looper
LOS	line of sight
LSI	large scale integration
MC	metering capstan
MLI	multilayer insulation
MTF	modulation transfer function
MRB	Material Review Board
MWFO	Midwest Field Office
NASA	National Aeronautics and Space Administration
NASTRAN	NASA Structural Analysis
NIIRS	National Imagery Interpretability Rating Scale
NPIC	National Photo Interpretation Center
NRO	National Reconnaissance Office
OAO	Orbiting Astronomical Observatory
OFTS	output fine tension sensor
OTD	Optical Technology Division
PDR	preliminary design review
PERT	program evaluation and review technique
PFA	post-flight analysis
PL	platen
PRB	Publications Review Board
PSQ	personnel security questionnaire
RCC	Recovery Control Center
RFP	Request for Proposal
RP	recycle profile
RTS	remote tracking station
RV	reentry vehicle
SALT	Strategic Arms Limitation Treaty
SCC	System Command and Control
SCF	Satellite Control Facility

SEM	scanning electron microscopy
SETS	System Engineering Technical Support
SH	shutter
SI	stellar index
SSM	sensor subassembly module
SSTC	sensor system test console
STC	Satellite Control Center
SV	satellite vehicle
SU	supply
SUA	supply assembly
TCA	two-camera assembly
TU	take-up
TUA	take-up assembly
ULE	ultra low expansion
UTB	ultra thin based
UUTB	ultra ultra thin base
VAFB	Vandenberg Air Force Base
V/h	vehicle's orbital velocity (V) over vehicle's altitude above Earth (h)
VP	velocity profile
WCFO	West Coast Field Office

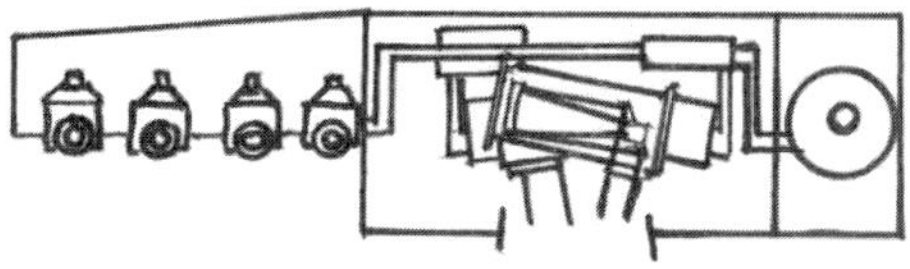

Chapter 1

Reconnaissance Cameras in Space

Phil Pressel

Photography from airplanes had shown its value during World War II, and during the early years of the Cold War the United States sought to develop new aircraft and new cameras to monitor developments in the Soviet Union. The story of the U-2 and the capture of Francis Gary Powers are well documented.

The Perkin-Elmer Corporation was involved in the design of a camera aboard the U-2. Photographs revealed the status of Soviet weaponry, discovering military assets and deployment. Reconnaissance photographs were a significant factor in the signing of the Strategic Arms Limitation Treaty (SALT) in the early 1970s because they provided a means of verification.

The Corona Program

The primary space orbital reconnaissance system in existence during the mid- to late-1960s and early 1970s was called Corona. Developed by the Itek Corporation, it was the most sophisticated space reconnaissance system at that time. The Corona camera was a film system that returned its film to Earth inside reentry vehicles.

The Corona program, the first orbiting reconnaissance satellite, began attempting launches in 1959. Like a true pioneer, that program broke new ground, endured failures, and learned many of the lessons that would be so successfully applied by Hexagon. Corona suffered fourteen failures (including one that nearly exploded on the launchpad) before achieving its first success in August 1960. Corona then began having successful flights. It continued in operation until 1972 when its mission was taken over by "Big Bird," the name given to Hexagon by the media people who became aware of a new, larger satellite being launched from Vandenberg Air Force Base since 1971.

Copyright © 2013 by the American Institute of Aeronautics and Astronautics, Inc., All rights reserved.

Real-time photographic solid-state technology was unknown during the 1960s and 1970s. Data transmission by electronic means was woefully inadequate then. Photographic information was transmitted to Earth by film capsules in reentry vehicles ejected from the satellite. After reentry through the atmosphere, a parachute was deployed and caught in midair by grappling hooks on an aircraft somewhere over the Pacific Ocean.

The early ground resolution on the Corona program was twenty feet. By the time the program ended in the early 1970s, various improvements resulted in ground resolution of around 6 feet, from approximately 100 miles altitude. This resolution meant that you could know that an object was around 6 feet in size, such as a car, but you could not detect any smaller details or smaller objects. The film recovered from Corona weighed between 10 to 20 pounds, and eventually went up to 80 pounds near the end of the program. Eighty pounds of film meant approximately 16,000 feet of film. The average duration of such a mission was about 19 days, with the satellite primarily operating at the beginning and end of the mission, with a long sleep time (sometimes called "zombie mode") in between.

In 1963, Corona was joined by another satellite system named Gambit. Whereas Corona was designed to photograph large swaths of territory at resolution good enough to indicate what newly detected objects were, Gambit had a more powerful camera that looked more closely at a much smaller amount of territory. These higher resolution photos provided more information about the objects that Corona had detected. The Gambit-1 system (also designated the KH-7) operated from 1963 until 1967 and had a resolution of approximately 2 feet. It was supplemented and then replaced by the Gambit-3 (or KH-8) system, which had even better resolution. Gambit-3 satellites operated until 1984.

The CIA and Itek

Itek was a small company formed in 1957 and was comprised of very competent scientists, engineers, and managers. Although Itek initially had excellent ties with the Central Intelligence Agency (CIA), over time this relationship began to deteriorate over development of a follow-on system to Corona. At the same time that the CIA and Itek were falling out, the relationship between the CIA, directed at that time by John McCone, and the National Reconnaissance Office (NRO), the organization responsible for coordinating satellite reconnaissance policy and directed at that time by Brockway McMillan, was also deteriorating.

During the mid-1960s, a new and a better photographic system was sought by the CIA's Directorate of Science and Technology (DS&T). Bob Kohler was a former Itek manager who developed the National Imagery Interpretability

Rating Scale (NIIRS) described in Chapter 2. Later he became a CIA technical analyst and eventually the deputy program manager for the Hexagon program. According to Kohler* the requirement came from John McCone, director of the CIA, to build a system that had the area coverage of Corona and the resolution of the KH-7. "Obviously what they were looking for was the ability to get the worldwide area coverage and get the fine things that you got with Corona but with enough quality that you could do better than Corona."

A panel consisting of people from industry and government, including the CIA, was formed to evaluate engineering approaches to improve resolution, coverage area, and increased film capacity. Among others, members of the panel included Edwin Land of the Polaroid-Land Camera Corporation and chairman of the President's Science Advisory Committee; two people from the Itek Corporation; and Rod Scott, a vice president and the chief scientist of the Perkin-Elmer Corporation. Scott was prominent in the development of spaceborne reconnaissance devices, in lightweight rugged optics and their mounts, and in new state-of-the-art image technology. The panel concluded that improvements to Corona would not enhance resolution sufficiently and that a totally new system was necessary.

The CIA's Challenge

The challenge was established in 1965 by CIA Director John McCone. He wanted a system that had the area coverage of the Corona program and the resolution of the KH-7. This meant the following:

- Improve the resolution of Corona (the predecessor program whose resolution was around 6 feet) by a factor of about three or more.
- Produce mapping capability to study most of Earth's landmass for a minimum of 39 days, but hopefully for months at a time.
- Periodically retrieve film capsules carrying a portion of the approximately 100,000 feet (about 20 miles) of film from each of two cameras, that is, 200,000 feet total.
- Produce photographs in stereo.

These requirements necessitated solving the following difficult problems (among many others):

- Avoid blurred photographs associated with a vehicle moving at over 17,000 miles per hour with a rotating camera containing film moving at 200 inches per second over the rotating Earth.

---

*Telephone interview with Bob Kohler in November 2005.

- Do not waste film.
- Control all of the mechanisms and film to very high speeds and accuracies to remain perfectly synchronized with extremely high reliability over long periods of time.
- Build a secured new structure in months—not years—to contain offices as well as huge test and assembly facilities.
- Design all the key motors in the system. There were no existing brushless motors that could run smoothly enough (without creating debris and "noise") as required for the job.
- Design sophisticated optics and mechanisms that remained aligned and functional after being subjected to a severe launch environment.
- Meet a rigorous schedule while designing most components from scratch.

ADDITIONAL REQUIREMENTS

The CIA's DS&T asked Itek to work with them in a study to develop a camera with better resolution over a much wider swath of Earth coverage. There were several specific goals of the new orbital camera system. The more specific requirements, which were more detailed than the original ones given by John McCone, were as follows:

- Improve resolution over the best that Corona could produce by a factor of approximately three, meaning between 2 to 3 feet, or hopefully better, at similar altitudes.
- Retrieve film capsules periodically, one at a time; each capsule containing significantly more film than Corona could carry. The goal was to carry a total of 200,000 feet of film for the mission (100,000 feet for each camera; that is 50,000 feet per each of four capsules).
- Produce both mapping and imagery photography over mission durations longer than one month and hopefully many months. Mapping capability would permit accurate wide-angle geographic studies of most of the landmasses of Earth, specifically the entire Sino-Soviet Bloc. Imagery would permit close-up studies of specific target areas of interest.
- Take stereo photographs in order to be able to determine heights of objects on the ground.

Photography from space presents many unique problems. Obviously for nonmanned vehicles, all of the operations had to be controlled remotely from Earth and repairs could not be made; therefore, reliability of all of the hardware and electronics had to be high.

DIFFICULTIES DESIGNING A CAMERA

There were many unique problems associated with designing a sophisticated high-resolution camera operating in outer space for missions lasting many months:

- How to avoid the problem of blurred photography associated with a "moving camera" and a moving target.
- How to scan Earth from side to side without wasting film.
- How to maintain the camera (originally aligned in a 1-*g* gravity field in the factory) in focus and in alignment after being subjected to severe launch *g* loads and vibration and then exposed to zero gravity and temperature extremes.
- How to keep moving parts in a harsh vacuum environment properly lubricated to minimize wear and tear over many months.
- How to control a complicated high-speed system to remain perfectly synchronized.
- How to test all of the components of the camera in all of its modes while simulating a space environment (except for zero gravity) in the laboratory.
- How to prevent the breakdown of certain materials in space and how to keep contamination (from the launch environment or from wear and tear) from floating around and redepositing itself in unwanted areas, such as right on the optics, or in ball bearings.

There are also a myriad of problems associated with handling of the film itself, such as

- How to keep the film flat at the focal plane.
- How to keep the film from wandering from side to side.
- How to keep the film edges from curling up.
- How to keep the film from being damaged or scratched during launch and operation.
- How to know what kind of environment the film needed to maintain its chemical and physical properties.
- How to know what amount of tension must be applied to the film to allow it to move smoothly along its path.
- How does the film make right-angle turns throughout its transport path in the vehicle and simultaneously move with the rotating focal plane while maintaining its own linear motion.
- How to control the high speed of the film.
- How to control electrostatic discharge on the film.

Of course there were also ample problems involved in the detail design of all of the electronics, mechanical parts, optics, and optical supports; the testing of the camera operation; and all of the logistics involved in fabrication as well as vendor, customer, and vehicle interfaces.

Most weather and communications satellites are placed in an equatorial orbit. These are launched from bases close to the equator such as from Florida. Note that Jules Verne was clever enough to come to the same conclusion in his book *Journey to the Moon and Back* where his rocket was launched from Florida.

Achieving a polar orbit from Florida requires the launch phase to proceed over land due to the Earth's rotation from west to east. This is dangerous, as the need to destroy an errant rocket would endanger anyone or anything under the rocket at the time of destruction. Therefore the satellite was to be launched from Vandenberg Air Force Base in California by the Titan III D rocket and eventually the Titan 34D rocket.

Just as the Hubble Space Telescope had the finest mirror ever fabricated, this book describes and explains how the finest film camera ever built came to be. The technical challenges combined with personal memories result in a marvelous story.

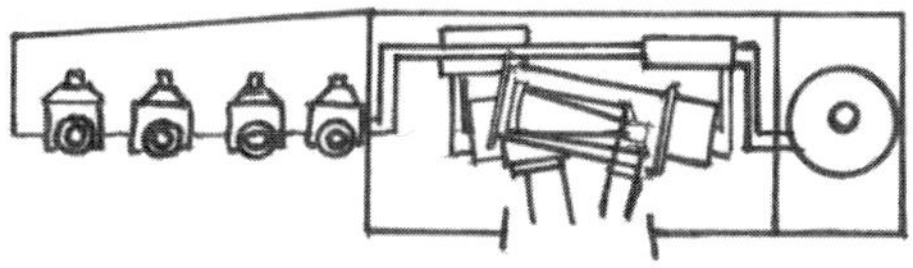

Chapter 2

Hexagon Camera System Overview

Phil Pressel

Cameras can be oriented two ways. One is crosswise to the vehicle's long axis dimension, and the other is along the vehicle's long dimension. One configuration was to rotate the camera first in one direction, then in the other in order to "see" on both sides of its path. Problems with this approach were that it would take time and power to stop/start/accelerate/decelerate the rotating camera. By the time this was accomplished, much of the territory on Earth could be well away from the field of view resulting in the waste of film during nonphotographic periods of camera operation.

Perkin-Elmer Enters the Picture

In the mid-1950s, as discussed in Chapter 1, Itek initially investigated camera configurations that would best meet the stringent requirements imposed by the CIA for a new reconnaissance system. Later Perkin-Elmer was asked to become involved in the project when difficulties were encountered between Itek and the CIA.

Perkin-Elmer's earliest participation in the Hexagon program effort began as a technology consultant to the CIA, in support of its ongoing effort with Itek and others, to develop a follow-on to the Corona reconnaissance satellite program. Within the company that effort was first designated the "Ad Hoc" program.

Perkin-Elmer's opportunity to propose a full system was thought by many within the company to be driven by the government's need for competitive proposals and they believed that we would, at best, play a supporting role to Itek after contract award. Fortunately for us at Perkin-Elmer, and for unexplained reasons, Itek appeared to take sides in a perceived competition between the CIA and the NRO for control of the next-generation reconnaissance system development effort and withdrew from the CIA-sponsored effort. As a result, the CIA vigorously sought an alternative contractor.

Copyright © 2013 by the American Institute of Aeronautics and Astronautics, Inc., All rights reserved.

Fig. 2.1 Dick Babish, Perkin-Elmer senior scientist.

According to Dick Babish (Fig. 2.1), a senior scientist and one of Perkin-Elmer's early leaders on the program, the CIA asked us in 1965 to perform a study for the camera. The study leader was Earl Brown, a pioneer in optics and space reconnaissance. Earl came up with a good comprehensive study encompassing parameters such as f-number, film speed, aperture, resolution, focal length, etc.

Perkin-Elmer came up with a completely different camera configuration than that of Itek. It was quite revolutionary and depended upon the development of three mechanical items, one of which was the brilliant invention by the late Don Cowles of our company. The device was called the *twister*. The second was called an *air bar* and was invented in 1963 by our director of research, Bob Hufnagel (Fig. 9.2). The air bar's shape, hole patterns, hole sizes, and hole shapes were derived to maintain film flatness and good tracking. The third item was the *looper* (see Chapter 7). These devices permitted recycling of the film so that none was wasted between frames and allowed the film to be rotated in synchronization with the optical bar while the film moved forward linearly. Essentially our concept consisted of rotating optics while Itek's consisted of a rotating camera.

During this study phase of the program, inventor Edwin Land visited our plant. He was on a presidential panel involved in intelligence and therefore a very influential person. He was also a visionary and loved technical innovations. When Don Cowles showed him how the twister worked with film speeding by, Land was blown away. I was impressed by Land's interest and sharp questions during a presentation that we made to him and other government officials. Certainly his opinions and desires for new reconnaissance camera systems helped in getting us the program.

The CIA transferred a large system mockup that was in development at Itek to us. Upon receipt, our personnel aggressively pursued bringing the mockup to a high level of functionality so that it might be utilized to successfully demonstrate film transport through a long and complex film path incorporating our signature air bar and twister concept.

As luck would have it, the mocked-up system functioned flawlessly on the day when CIA management and program staff personnel observed its performance. It wasn't functioning correctly the day before that visit, nor was it functioning the day after that visit. In fact, it was quite a few weeks before the demonstrated level of performance was duplicated. The CIA folks were astute enough after witnessing the scheduled morning demonstration to request an impromptu demonstration later the same day. Had they waited a day before requesting the second demonstration, the Hexagon program history might have played out differently.

Our study was presented to the CIA at night in an innocuous-looking safe house in Washington, D.C. Albert "Bud" Wheelon, the first CIA Deputy Director of Science and Technology (from 1963 to 1966), told us that they thought highly of our concept. We assured them that we could do the job and this gave us the opportunity to write a proposal to bid on the job.

Preparing the Proposal

In the spring of 1966 we received the formal Request for Proposal (RFP). In Chapter 5, my colleague Sam Palasciano describes the technical requirements, how they were fulfilled, and the CIA's review of the proposal. It all took place quickly—the Ad Hoc project had six weeks to solidify the overall design and have enough analyses and systems studies to prove that our concept was feasible and achievable. By now Perkin-Elmer had a well-staffed drafting room that drew all of the major components and assemblies on large paper drawings. We all worked long pressure-filled days.

Monitoring everything was our leader Mike Maguire (Fig. 2.7). He was a smart, tough, and inspirational leader and was respected by all of us. Technically he knew the engineering workings of the program inside out. He and his planning and financial team worked with the Engineering department to plan out reasonable schedules and milestones. The Purchasing and Manufacturing departments were busy contacting vendors to get quotes for the fabrication of major components.

The proposal phase was incredibly hectic. There were approximately 30 of us who were cleared on the program. During the study phase we had come up with preliminary designs and sizing for the major assemblies of the system, but there were many unknowns that would not be defined until the program would be well underway. We had to make many assumptions. Fortunately there was excellent leadership and communications between various departments and between the engineers responsible for the major assemblies and electronic controls. We performed critical analyses and altered the designs as we proceeded in order to meet the customer requirements and our own trade-offs.

The actual proposal writing process became like a well-lubricated machine. We would write by hand (we did not have computers or word processors in 1966) or on a typewriter and give those writings to the department secretary to put into the proposal format. The head of Engineering was Bob Landsman (Fig. 2.7), an electrical engineer and knowledgeable in everything. I also worked with Mike Krim, the mechanical engineer who hired me. He was a creative engineer and a wonderful freehand drawing artist.

Our secretary at the time was Millie Roman, a charming lady. While she typed, some of us would speak into a Dictaphone for her to transcribe later. Everyone's writings then went to our department manager for review (if he had time) and then to the primary two people who put the proposal together. The two were John Garrish, an opto-mechanical/systems engineer, and Earl Brown, one of the chief scientists of the company and a well-known expert in optics and reconnaissance as well as an author of noted reference books.

The two of them would review and edit the material, then cut and paste using actual scissors and glue or tape to put it all together. There were several very large sections to the proposal. They ranged from technical descriptions of electronics, optics, mechanics, system performance predictions, assembly flow and testing, to administration, finances, and planning.

The mechanical engineering volume was at least 1-inch thick and written primarily by Mike Krim and me. It contained drawings created by several draftsmen, including Ed Brelling, Len Petruzzelli, and Dick Chester. I continued to work with them for many years on the program. Len used to love to bet on the horses and each summer spent vacation time at Saratoga racetrack in New York. Ed was very free with comments and suggestions on various designs; he was usually right and came up with some clever ones. Dick was a longtime employee and had managed the Drafting department before this program came along.

The Contract is Awarded

Perkin-Elmer won the contract. Our management recognized that a post-award replanning of the Hexagon program schedule was in both the government's and the company's best interest. To the surprise of the assigned staff, Perkin-Elmer's president, Chester Nimitz Jr., called for that replanning in a meeting held shortly after contract award.

In the early meetings, management set forth the suggestion that program success could be best assured in an environment of open and constructive confrontation on all technical and business matters with the customer and associated consultants. It was felt that if each party assertively proffered its thoughts on what was needed to achieve program success, then reasonable compromises would lead to a sensor system that met the government's mission requirements. It is a testimony to the character of all involved that the ensuing open dialogue built a foundation of cooperation and innovation between the CIA and its contractors. This communication served the interests of all parties in ensuring a record of mission successes, beginning with the very first Hexagon flight.

Staffing the Program

Assembly of a skilled technical staff to start the program was accomplished in an environment where the staff recruitment process was encumbered by the inability to reveal the nature of the program due to security considerations and the specific task to which any candidate would be assigned. In addition, the time it took to complete required background security checks was long (from 4 months to a year) and ultimate employment was contingent upon surviving the security screening process.

This meant that the company had to immediately recruit many hundreds of employees to staff the program. As new people came they were briefed and assigned to work with some of us old-timers. Prior to receiving their clearance most of the newly hired staff worked in a nonsecure bullpen on nonclassified work. That area of the building was known as "the tank"—also known as "the mushroom patch" because people in there were kept in the dark and fed a lot of manure.

The main engineering room had dozens of desks next to each other separated by aisles. There were several private offices around the periphery for the managers of Drafting, Mechanical, Electrical, and Systems Engineering. The whole set-up continually changed as new staff members came onboard. The crucial thing was that all employees in that facility were cleared and dedicated to this one program and the whole area was easy to secure. All the windows had been covered with black material. We could all discuss the program openly without fear of being overheard by people without security clearance.

Some of us who were already cleared for the program were asked to help interview a number of applicants, including senior management people for whom we would eventually work, to recommend hiring someone or not. Some background information had to be obtained in order to get some feel for whether the person had a reasonable chance of passing his security investigation. I suppose that some of the questions that we asked in those days would not be allowed today due to privacy invasion limitations.

One of the people that I interviewed for the new program was Charlie Karatzas, who would soon become my boss and one of the best mechanical engineers I have ever known. I found out about his vast knowledge when he first applied for a job. I had come up with some clever questions dealing with mechanism design, optical mounting, and knowledge of materials. Charlie knew a lot more than I did. He was soon hired and appointed manager of the Mechanical Engineering department. I was really happy about that. For years I had the privilege and good fortune to work with Charlie, learn from him, and enjoy his stories and sense of humor.

Immediately after we won the program, Bob Jones (Fig. 4.4), a systems and mechanical engineer, was appointed director of the Engineering department. He assigned some key people to be project engineers for the major subassemblies. He appointed me to be the project engineer on the optical bar, the actual optical camera (Chapter 8). This meant that I would be responsible for its mechanical design and would be the lead interface to the other engineering disciplines such as electrical, systems, manufacturing, reliability, quality, and test. I would also lead design reviews for our customer, the CIA.

The whole Hexagon program involved many other companies besides Perkin-Elmer. They included Lockheed (now Lockheed Martin), McDonnell Douglas, Kodak, Itek, General Electric; consultant companies such as TRW and The Aerospace Corporation; plus the government, including the military.

What made our company Perkin-Elmer so successful was its outstanding managers and scientific and technical staff. Over the years they had some world-class optical designers and optical scientists. It is impossible to acknowledge by name the many hundreds of employees who contributed to the Hexagon program including engineers, draftsmen, technicians, and associated staff. Needless to say, without them, success would not have been possible.

The Hexagon Camera System

To better understand the Hexagon system, let's first compare it to the most famous camera ever launched into space, the Hubble Space Telescope. The

Hubble and the Hexagon systems had several important differences. The Hubble orbits Earth and looks at the stars. The Hexagon orbited Earth and looked at it. The Hubble optical design was completely different. It transmits photos in real or near-real time using solid-state electronic devices rather than film and operates in visible, infrared, and other spectral ranges. The Hubble is basically a static structure with few moving parts. The Hexagon camera returned its photos in film capsules, and the photos were in the visual spectrum. Its major components contained many high-speed moving parts. It contained hundreds of bearings, large and heavy rotating assemblies, and film moving rapidly in linear and rotational modes, all of which had to be accurately controlled electronically. Even today the Hexagon camera probably still remains the most complicated system ever put in orbit.

The Hexagon camera system (Fig. 2.2), was designed, assembled, and tested at our plant in Danbury, Connecticut then shipped to Lockheed (now Lockheed-Martin) in Sunnyvale, California to be integrated into the vehicle and undergo further thermal vacuum and other tests.

CONFIGURATION

A schematic drawing of the basic vehicle is shown in Fig. 2.3. It consisted of four take-up assemblies contained within four reentry vehicles located in

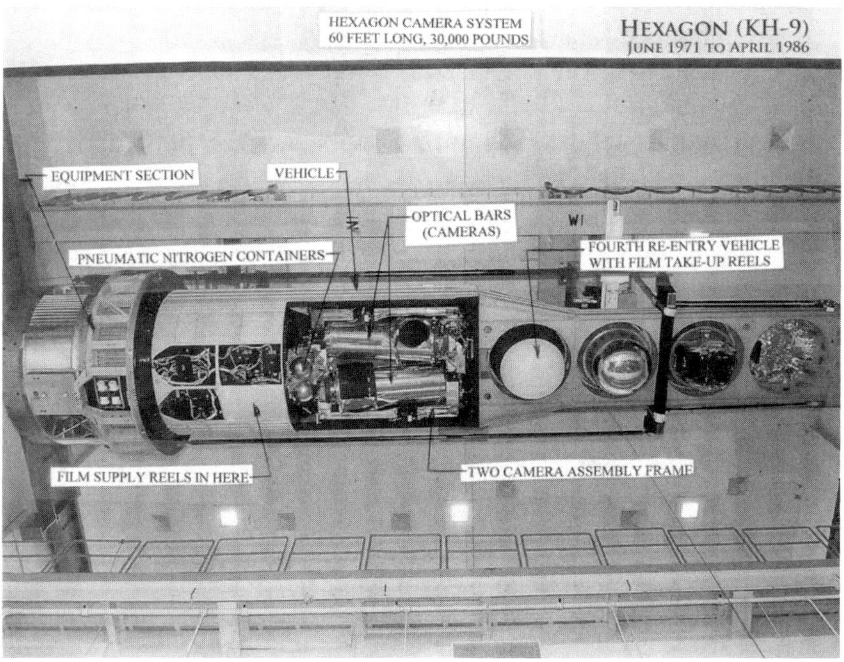

Fig. 2.2 The Hexagon camera system.

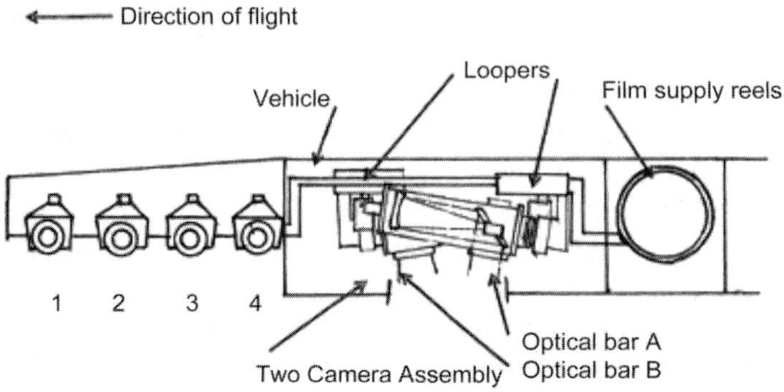

Fig. 2.3 Overall vehicle configuration of the Hexagon camera.

the forward nose cone section of the vehicle. The cameras were located in the midsection of the vehicle, and the film supply assemblies were contained in the aft section. Most of the camera-control electronic boxes as well as the vehicle's electronic, attitude, and power controls were located in the equipment section found behind the aft section.

There were two identical camera systems in order to provide stereo coverage. Also if one camera failed, the other one would still be able to support the mission. Each camera system had its own film supply, take-up, and film path. The cameras were contained in an assembly called the two-camera assembly (TCA) shown in Fig. 2.4.

It is important to point out that the supply and take-ups were totally separated structurally from the rest of the TCA, except for being connected by a nonstructural, delicate (wimpy) component—the film. Early in the

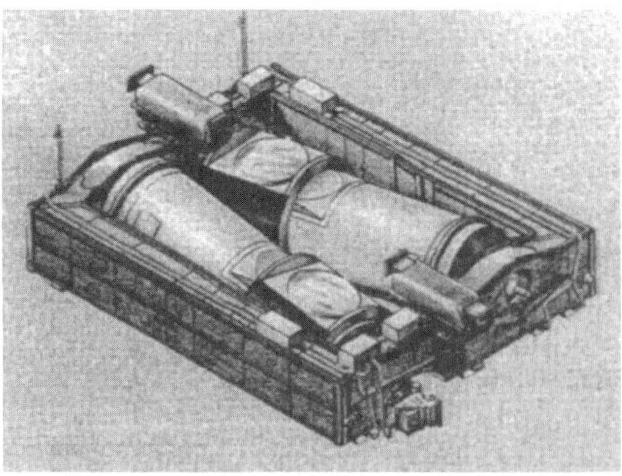

Fig. 2.4 TCA isometric.

development of the program, Lockheed personnel working with Perkin Elmer mechanical engineer Mike Krim decided that it would be advantageous to separate the TCA from everything else in order to prevent any extraneous vehicle deformations during launch or orbit. They recommended that the TCA be attached to the vehicle in a kinematic manner with three-point supports in order to define a single plane of supporting structure.

This was accomplished by tying the aft part of the TCA frame at two places directly to the vehicle and attaching two pitch links at the forward end going from the vehicle to the frame. The two vertical struts had spherical bearings and a ball-screw pin connection at the end of each strut. A simplified explanation of the ball-screw assembly is that it consists of a long screw approximately 0.5-inch diameter and a set of bearing balls (about a 0.25-inch diameter). The screw was rotated by a motor thus driving a pin linearly in or out of its receptacle. One pitch link could be decoupled remotely on orbit to provide a three-point mount.

Decoupling was accomplished by disconnecting one ball-screw pin connection. Once during a test, one of the ball screws seemed to be too tight and made a grinding noise. The technician and inspector working the job took the affected ball screw apart. Using a micrometer, they began to measure the diameter of all 128 balls in the ball screw. They found 127 balls of the exact same diameter and one that was 0.001-inch oversize. The 127 "good" balls were placed in one tray and the oversized ball in another tray. They called the engineer down to the floor to show him what they had found. The engineer picked up the bad ball and accidentally dropped it into the tray with the "good" balls. The technician had to start measuring all the balls again. He was not happy.

The supply and take-up reels rotated in opposite directions in order to cancel out angular momentum. The optical bars also rotated in opposite directions for the same reason as did all heavy rotating masses.

The Film

The optical bars each performed a whiskbroom scan of Earth as the satellite moved forward. As each optical bar rotated, it successively imaged narrow slits of ground imagery onto the synchronously moving film. During one frame of exposure, a swath of the ground was captured. The swath measured 10 nautical miles wide in the vehicle-in-track direction and up to 170 nautical miles in the cross-track direction (at 100 nautical miles altitude).

The optical bars rotated continuously in their respective directions at constant velocity depending on V/h (the orbital velocity V of the vehicle divided by the vehicle's altitude h above Earth). Photographic imaging occurred only during scans of +/−60 degrees or less on either side of *nadir* (camera pointed straight down), depending upon what images were required for intelligence purposes. After retrieval and development of the film, photo-interpreters could look at certain narrow areas of the wide-angle photos to obtain very detailed information. It was during these scan periods that the film's linear velocity and rotational speed in the platen had to be synchronized exactly with the image in the optical bar.

If the film were kept moving forward during the nonimaging portion of the optical bar rotation, there would be an unacceptable waste of film. In fact, if film ran at constant velocity, 83 to 96 percent of the film would be wasted. Similarly, film could not start and stop during the optical bar nonphotographic periods because this would require infinite (or at least very high) accelerations and decelerations of the film spools. This was not practical and would have required a great deal of power.

The film exited the supply reel at a constant velocity of 70 inches per second. After the film left the supply, it had to be moved in accordance with a prescribed film velocity profile to enable photography to occur at the proper time and to utilize as much of the film as possible.

The film was accelerated to photographic speed in the platen. At *perigee* (the lowest point in the satellite's orbit), this was 204 inches per second at the focal plane. After exposure occurred, the film was then decelerated and driven backwards so that the next exposure was made with only 2 inches of film between exposures. The film was then stopped so that an electronic data block could be inscribed on the film in this narrow space. At altitudes higher than perigee all of the film speeds slowed down proportionately.

The film was fed through and controlled in a device called a platen. The nonrotating part of the platen was attached to the nonrotating part of the optical bar and its rotating/oscillating portion was synchronized to the rotating portion of the optical bar. The real key to the success of the Hexagon camera system was the invention of the twister. The twister, containing air bars, allowed the film to travel both linearly past the focal plane slit, while being rotated in synch with the rotating focal plane.

The job of accommodating the film velocity profile from constant velocity at the supply to a variable at the focal plane in the platen and storing the film during the nonphotographic cycle of the optical bar was accomplished by means of a film storage device called the looper. The proper quantity of film was fed into the "entrance" portion of the looper prior to being exposed in the platen.

After exposure, the film was stored in the "exit" side of the looper. It was then wound up at constant velocity again 70 inches per second at perigee onto the take-up assembly in the forward section of the vehicle. After the first of four take-up reels (each in its own reentry vehicle) was filled, the film was wound and cinched onto the core of the next take-up reel and then the film was cut. At the appropriate time during one of the following orbits, the filled reentry vehicle was jettisoned and returned to Earth.

The entire process was repeated until all of the film was used and the fourth reentry vehicle returned to Earth. The remaining Hexagon vehicle was now useless and left to eventually be deorbited and burnt up during reentry to Earth. Unfortunately, it could not be reused or retrieved.

The System is Launched

The first system was launched on June 15, 1971 after about five years of development, and the nineteenth and final operational system on June 25, 1984. Each of them was an outstanding success. The launch dates and parameters for each launch are in Appendix A. The last camera was launched on April 18, 1986, but the launch vehicle exploded at an altitude of 800 feet 8 s after liftoff (see Chapter 19)—3 months after the Space Shuttle *Challenger* disaster. We had all been so proud of each Hexagon flight and its accomplishments that upon hearing of the explosion minutes after it happened, many of us in the plant cried. I was overcome by a feeling of terrible loss. It was our baby, and now after all of those successes, the last one in the production line was destroyed.

The basic configuration of the system's main components did not change much from start to finish; however over time many improvements were made, including the following:

- Introducing lengths of color film to the black and white film
- Obtaining thinner film, thereby increasing the film capacity of each reel of film from 100,000 feet to approximately 150,000 feet
- Building a larger "super looper" dynamic film storage device to allow more operational flexibility, allowing photography over small areas without starting up supply and take-up, a relatively long procedure
- Adding a star-tracker camera in the early 1980s to improve the knowledge of exactly where on Earth the cameras were taking photos
- Adding a remotely replaceable optical filter in the camera during 1973
- Substituting newer solid-state electronic devices, thereby increasing the reliability of the system and time in-orbit

All of these improvements resulted in photographs with resolution better than the original requirements, and the mission orbital times increased from the original 52-day mission to missions lasting as long as 275 days.

The Hexagon camera (Fig. 2.5) had the capability of seeing relatively small things on the ground. These numbers may be exaggerated but excellent

Fig. 2.5 Hexagon KH-9 vehicle.

resolution was indeed achieved for certain orbits. Some missions had reentry vehicles sent back sooner than originally planned due to whatever world crisis was going on at the time so that photo-interpreters could see crucial territory.

In addition to filming intercontinental ballistic missile (ICBM) silos, airfields, Soviet shipyards, and other strategic weapons sites, the Hexagon also did more esoteric work as well. There was a special task force for Earth resources. Later Hexagons could carry special films for gathering information on crops and agriculture that could be used for economic intelligence. If the Soviet Union had a bad wheat harvest, it could give the United States added leverage. The system could also help with national disasters—a Hexagon satellite was used to look at flood plains during one of the Mississippi spillovers. Indeed, Hexagon was used for many special things that the civilian spacecraft Landsat could not do because it didn't have the required resolution.

For all of the missions, out of 76 reentry vehicles, only one was lost. Apparently due to a parachute failure on the reentry vehicle, the airplane that was supposed to catch it using a grappling hook was unable to do so and the reentry vehicle sank over 16,000 feet into the Pacific Ocean. Chapter 18 contains a description of the design of a rescue device and the attempts to recover the vehicle and the film.

The vehicle was not reusable. Late in the program's existence (early 1980s) I was assigned to a team to determine if it was feasible to put the last system on a pallet on the Space Shuttle so that it could be flown and then reused, rather than discarded. This was considered very secret at the time. I worked in an area that was like a vault with Arnie Wallace and Larry Cassidy behind closed doors. I came up with some drawings for a pallet and showed that even with the two new S-cubed (mapping cameras) mounted on the frame, it was indeed feasible. Besides using the camera for Earth reconnaissance, the customer wanted to send it to orbit and look at Mars. Too bad it was never done. It could have been done so cheaply at the time. I don't know what happened to the idea.

Success of the System

The public was not aware how immensely successful this program was. Many of us, however, received the award shown in Fig. 2.6. Several items relating to the award are quite meaningful. The shape is six-sided and therefore coincided with the program's code name "Hexagon." The insert in the glass itself was made of film and showed 20 stars. Each star signified a vehicle that was launched including the final one that exploded during launch. The map indicates that photography covered the whole Earth and the inscription denoted the timeframe 1971–1986 and our motto for success, "We Met the Challenge."

Our original general manager and program director was Mike Maguire. His importance to the program is discussed in Chapter 4. A photograph of him taken during retirement, alongside two other important members of our staff, is shown in Fig. 2.7. The others are Bob Landsman, our first engineering manager, and Dick Carricato, a servo engineer. Dick Carricato was familiar with the entire system and developed the equations that governed the image motion control system. He was a key person: he was

Fig. 2.6 Award for the Hexagon program.

Fig. 2.7 Bob Landsman, Mike Maguire, and Dick Carricato.

usually the first one asked to solve difficult systems problems, especially ones that developed in-orbit when the only information available was the diagnostic data transmitted. After the successful conclusion of the first mission in 1971, Mike received three letters of commendation. They are shown in Figs. 2.8, 2.9, and 2.10.

Several former officials of the CIA were interviewed for this book. What follows are some quotes from them concerning the importance of the Hexagon program. According to Bob Kohler,* a former CIA deputy program manager on Hexagon, "Certainly this program found a lot of stuff both in the Soviet Union and China. I think that mobile missiles were also found. There were some underground facilities that this program found. It became the vehicle for SALT I verification."

Kohler also discussed the now-famous NIIS that he co-developed, saying

> It was first developed for this program. It was a number system, and it is still in existence. It goes from 1 to 9; for each number it gives examples of intelligence of things that you can do with that quality. Corona tended to be 2 and 3, and this program tended to be 3, 4, and 5. It was developed specifically for the purpose of SALT verification. The intelligence community wanted to know if the quality of the pictures was good enough to verify if the Soviet Union was doing something. I believe the magic number was 4. If it got a rating of 4 or higher you could have confidence that they were abiding in SALT I. If it was not 4 then you

*Telephone interview with Bob Kohler, November 2005.

THE PERKIN-ELMER CORPORATION NORWALK, CONNECTICUT, U.S.A.

ROBERT H. SORENSEN SENIOR VICE PRESIDENT, OPTICAL GROUP

July 21, 1971

Dear Mike:

At our Perkin-Elmer Board Meeting on Thursday, July 15, the following resolution was passed by our outside directors:

"RESOLVED, that Messrs. Ammidon, Dacey, Hammond, Lewis, Swain, Thomas and Townes, being all of the directors of this Corporation who are not officers thereof, wish to express their sincere admiration and appreciation both as directors and as citizens to the management and those employees of the Corporation concerned with the outstanding technical achievements in the national interest reported to this meeting in connection with the performance of certain Government contracts; and

FURTHER RESOLVED, that the tenor of the aforesaid Resolution be conveyed to all individuals concerned."

May I add for Mr. Nimitz and myself our additional congratulations for the successful events of the last thirty days, made possible by the professional execution of a difficult task over several years of personal dedication by you and your people.

Best regards,

RH Sorensen

Mr. Michael F. Maguire
Optical Technology Division
Wooster Heights Road
Danbury, Connecticut 06810

Fig. 2.8 Letter from Bob Sorenson, Perkin-Elmer Senior Vice President (later he became president).

> didn't have confidence. That scale is used now for every system for all kinds of stuff.

When asked if this program was crucial to the government, Kohler answered, "It was certainly crucial for SALT verification. No question about it. It was the way. Nothing else had the area coverage. It became what was known as national technical means of verification."

24 August 1971

Dear Mike,

I would like to express my deep appreciation for the honors that were extended to me last Thursday and for the thoughtfulness in inviting Dorothy to share the evening with me.

As with all relationships there comes a time for change and it is appropriate that I move out at this time. Our relationship over this period will not soon be forgotten and I will always remember you as a fair and competent compatriot and competitor. It has been a privilege and a pleasure to work with you over the past years despite the many problems and controversies which occurred. Your leadership in the system development I am certain was the greatest single factor leading to the success which has been achieved.

I am sure you will find Don a very competent and fair person with whom to deal, and I feel satisfied that the program is in good hands. I hope that the relationship between our respective organizations will continue to improve and reach the level of mutual trust and confidence I feel it deserves.

Would you please extend my sincere thanks and appreciation to all the people who have worked hard and sacrificially to make the program a success. I would like to write to them all personally but I'm the world's greatest procrastinator at letter writing.

I wish you the best of luck for the future.

Sincerely,

Don

Fig. 2.9 Letter from Don Patterson, first program manager for the CIA.

John McMahon,[†] former deputy director of the CIA and of the special projects office that included all satellite activities within the CIA DS&T, and also former president of Lockheed, said that the Hexagon satellites "filled in a span of ignorance that we possessed. But the key was that they permitted us to see all the airfields at once and we could count airplanes."

[†]Telephone interview with John McMahon, October 26, 2005.

5 AUG 197

Mr. Michael F. Maguire
Vice President and General Manager
Optical Division, East
Perkin-Elmer Corporation
Danbury, Connecticut

Dear Mike:

I attempted to reach you earlier but you were away either on a trip or enjoying the luxury of some leave; time has passed further so that I am now extremely remiss in extending my congratulations to you and the Perkin-Elmer Team. It is well known that success has many fathers, but I firmly believe that you hold the sole claim as the legitimate one. I trust you will take full measure of pride in what you have accomplished and the contribution which you have provided our Nation.

I was sorry to miss your celebration here in Washington, but a complicated schedule became even worse and I found no way to participate in something which I sincerely desired to attend.

Please extend to Chester and Bob Sorenson my words of congratulations.

Sincerely,

John N. McMahon

Fig. 2.10 Letter from John McMahon, Deputy Director of Special Projects, CIA DS&T.

Roy Burks,[‡] former CIA technical director of the Corona and subsequently CIA/NRO manager of Hexagon technical interfaces, was very involved in the initial phases of Hexagon and was also in charge of field integration and test for Hexagon on the West Coast in the late 1960s and early 1970s (Fig. 2.11). "Without that means of technical verification by overhead systems, SALT would not have occurred," Burks said. Burks also felt that the relationship between all the key organizations involved

[‡]Telephone interview with Roy Burks, October 16, 2005.

Fig. 2.11 Roy Burks, CIA/NRO manager of Hexagon technical interfaces.

in the program was excellent, even when managerial control was passed from the CIA to the Air Force in 1972: "In fact when the program was transferred to the Air Force, they asked me would I come and assist them for the contract negotiations. I did so and we had no real problems." Burks explained, "At that point the program was operational and the agency was tied up with the real-time system. The Air Force was also in charge of the launches and it was decided that it would be better for the Air Force to manage the system and all of the contracts that were on the West Coast."

The phenomenal success that Hexagon achieved was a result of teamwork by a talented group of people. This organization is described in the next chapter.

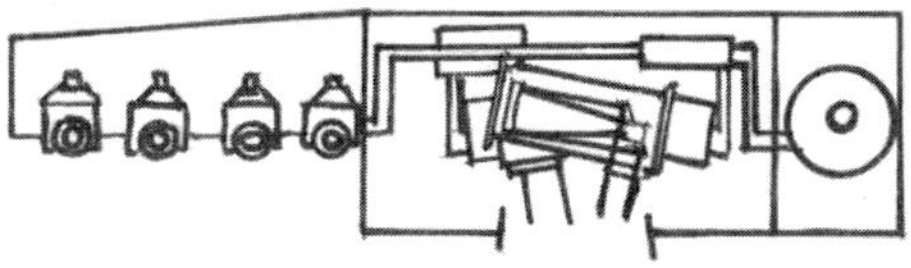

Chapter 3

The Perkin-Elmer Organization

Phil Pressel

The Perkin-Elmer Corporation began in 1937. It was created by 30-year-old investment banker Richard Perkin and 64-year-old publisher and court stenographer Charles Elmer. They were both enthusiastic amateur astronomers. They agreed that it was a pity that the United States had no significant producers of fine optics and astronomical instruments. They pooled some start-up funds, and from the beginning the company was successful. During World War II the company became the main source of precision optics for use in range finders, bombsights, tank periscopes, and aerial reconnaissance lenses.

The company developed sophisticated ground-, aerial-, and space-based telescopes and cameras, in addition to becoming a world leader in the development of analytical instrumentation such as gas chromatographs for universities, hospitals, and the chemical industry. Much of the work involving optical components, lenses, and electro-optical systems was done for its major customer, the U.S. government, including NASA and the military—much of it classified. Some of the telescopes that the company designed and built were the Stratoscope I and II telescopes suspended on a balloon, the Orbiting Astronomical Observatory (OAO), the Solar Telescope for Spacelab, and later during the 1980s the Hubble Space Telescope for NASA followed by the AXAF Chandra X-Ray telescope.

In 1969 Fortune magazine reported that the company was one of the largest in the United States with sales of $200 million and over 7000 employees. In the mid-2000s, the company merged with biotechnology companies and refocused on health sciences

While I was working there our CEO was Chester Nimitz, Jr., son of the famous World War II fleet admiral. He too was an admiral, commanding submarines: commander of the Haddo (SS-255), a Gato-class submarine that was commissioned in 1942 and ended its service in 1946. He led the submarine in seven war patrols in the Pacific. Overall it sank a Japanese frigate, a destroyer, a minelayer, merchant cargo ships, tankers, and trawlers.

Copyright © 2013 by the American Institute of Aeronautics and Astronautics, Inc., All rights reserved.

Chester Nimitz Jr. went onto an executive management career at Texas Instruments Corporation prior to coming to Perkin-Elmer. The president of Perkin-Elmer during these years was Bob Sorenson.

In 1981 sales reached a billion dollars and the company was named one of the five best managed companies in the United States by Dun's Review. Unknown to the public, a major reason for this outstanding growth in the 1960s and 1970s was the contract for a highly classified and successful spy-in-the-sky Hexagon camera designed and built by us. This work started in 1965 with a study contract and culminated in a multimillion dollar contract in October 1966. This contract was extended several times over the next decade and a half for a substantial amount of money.

Our Facility

The company needed to conceal the scope and objective of the Hexagon program due to stringent security requirements. That led to the adoption of a cover story that a number of separate restricted-access programs were being executed under the umbrella of an organizational entity known as "Programs 1." The Programs 1 staff was destined to grow to 1200 people in just 15 months and a new secure facility was created 15 miles to the north. In April 1968, approximately a year and a half after we won the job, the whole program was moved from our temporary headquarters in Wilton, Connecticut about 15 miles north to the new facility in Danbury.

The new building housed all of the department offices necessary to run the program. It also contained an assembly area and a test facility for optical, thermal-vacuum, vibration, and acoustic testing. That facility was completed at the company's expense and on company-owned property. The government provided much of the test equipment, including two huge thermal-vacuum chambers, one of which was one of the largest spherical chamber in the world at that time.

The office space in our new building was a great improvement over the initial days spent in our previous building where we performed the study and wrote the proposal. The old area had been isolated from the rest of the company working in the same building. All of the windows had been covered with black plastic and most of us worked in one of several large rooms that contained a dozen or so desks and some drafting tables.

I remember the first day that I drove to the new facility. It was in early April 1968; it was very foggy that morning. As I went up the company driveway, this large white structure loomed ahead. There were no windows. It looked bleak and gloomy and like a prison through the fog. The only windows in the building at that time were in the cafeteria. I was to spend 27 happy and productive years, the rest of my career in that building (Fig. 3.1).

Fig. 3.1 Home of the Hexagon project (photograph taken 1968).

The whole building was dedicated to Hexagon. Construction was still going on in the assembly areas of the building, but most of the offices were complete. I found my way to the mechanical engineering area located on the second floor in the middle of the building. It was quite nice. We were located near the electrical and systems engineering areas. The drafting room was just across the hall and contained approximately 15 drafting boards. All of the managers had offices on the periphery of the room. The second floor also contained the senior executives and the Quality, Reliability, Assembly, Test, and Research departments.

The center of the large room was divided into many cubicles that could seat two or four people. The carpeting was a gray-brown color. The walls of each cubicle were yellow with vertical black stripes outlining the edges and corners. Each person had the traditional gray metallic desk. Above the top left- and right-hand drawer was a pull-out shelf for extra working area. Some of us put our feet up on these shelves when deep in thought or when talking on the phone.

Because all of the work was classified as top secret most of us engineers, as well as the senior staff and managers, had safes. As the program progressed, more and more safes appeared. I wound up with two safes to store all of my documents. These were big and heavy safes, each weighing approximately 1000 pounds, about 5 feet high with four drawers. There was a log book attached to each safe into which you signed in and out every time you opened and closed it. You were not allowed to keep a safe open except to retrieve or replace a document. It did become a nuisance when needing different documents during the day. If a safe was left open overnight, you received a violation and after three violations you were threatened with dismissal. Needless to say that I made sure (by leaving myself notes in various places) to always close them.

One of our new mechanical engineers with security clearance was of Russian background. His name was Alex Siziew, and he spoke loudly with a strong Russian accent. Here we were doing secret work to spy on the Soviet Union, our primary Cold War enemy, and we could hear him on the phone speaking Russian to his wife. It could be quite disconcerting. He was a very nice person and a competent engineer who designed several important tests for us.

The first floor had the assembly areas, test areas, machine shop, optical shops, and administrative department. The only entrances to the building were at the front lobby, one side of the building, and the rear of the building. A guard manned each entrance.

Soon after the opening of the building a medical area was created and staffed with our first nurse, Elizabeth O'Connor. Years later she was succeeded by Pat McWhirt. Both were wonderful and became friends with many of us. About a year after we moved in, a barbershop was established in the building. This was to service the busy men in the building (there were basically no women engineers or technicians in those days) and save time as most worked late hours. The barber's name was Bothelo and he was in the building two days a week. His regular barbershop, as of 2009, was still on South Street.

The cafeteria was on the second floor in the front of the building. As mentioned, it was the only area that had windows. It was a pleasant place to be and had good food. It was also the site for many parties. After each major design and system review with our customer and associated contractors, there was a late afternoon or evening dinner party with shrimp cocktails and other appetizers with a choice of alcoholic beverages. Dinner was usually roast beef. These parties were for the company staff that had contributed to the subject discussed. After some of these meetings, particularly dealing with final design reviews of a major assembly, the general manager or the director of Engineering sometimes invited the participants to a restaurant.

The major conference room was a beautiful modern wood-paneled facility. It contained a large glass screen with two viewgraph machines located behind the screen. Someone placed the transparent overhead slides on the machines as directed by the speaker. Sometimes there was a problem with the speaker up front wanting to show slides out of order—another person ran the slide projector in back of the room. The computer projection system was still decades from invention.

There was a long conference table extending from side to side at the front of the room where large blueprints could be placed. There were approximately 40 upholstered armchairs for the audience. There was an American flag in the left front corner of the room behind a podium.

Design reviews were held periodically for the major assemblies, such as the supply, looper, platen, optical bar, or take-up, as well as for overall system evaluation. The major presenters were the mechanical, electrical, and systems engineers. The systems engineers presented the main requirements; predicted system performance; and results of thermal, stress, and dynamics analyses.

The senior customer and our people usually sat in the front. Most of the meetings went very well. However sometimes there were heated discussions between our company personnel and the customer or some of their associated contractors, some of whom were contentious. Some of the heated debates concerned anomalies in test results, data interpretation, and program plans.

There was another critically important room. It was located on the second floor near the executive and vehicle managers' offices. This was called the "war room." It was a large rectangular room with special wallboards surrounding the room with its large conference table. The rear wall of the room was covered with a map of the world and a sliding transparent screen containing a line describing the passage of the orbiting satellite. The slide could be placed over wherever on Earth the satellite was above.

The remaining walls were covered with orbital data coming from the cameras. These included charts, graph plots, and all transmitted diagnostics. The construction of the room included a special wire-mesh screen embedded behind the walls to prevent any interference or eavesdropping. There were also special secure telephones and data links to other installations. This was an extremely secure room. The war room was used every morning at 7:30 where key people gathered to review: first, progress on our schedule charts; then progress and problems for each model, starting with the engineering model; and finally the flight models. When a satellite was in-orbit, there were daily morning meetings to go over data and plans. Later during the project, when the flights were so far between, the war room was closed much of the time unless there was a launch or an on-orbit problem.

The telephone system for the majority of the building was unique. There was a phone on each person's desk. The handheld receiver had a switch in the handle that needed to be depressed when talking. This was to provide extra security for telephone conversations. There was also an area with scramble phones. The most effective security precaution for the phone system was to avoid talking about classified things. When we did talk about classified information, it was necessary to use code names for key items.

OUR STAFF

The leadership started with Mike Maguire, our general manager, vice president, and program director (Fig. 2.7). Mike was indefatigable, smart, and an assertive and motivating manager. We all worked at least 60 h per week and Mike was always the first one in and the last one out. Discussions between many of us during the program and in even retirement always resulted in us saying in one way or another that "Mike was the best manager I ever worked for."

When I asked him about the program well into our retirement years, Mike told me, "This was a career job. It had to be for most of us. How many people get a chance to build something that sophisticated, and that did so much for the country? It was a marvelous opportunity to do something really spectacular and really well." He also agreed with me that this was probably the most complicated thing ever put in-orbit. He said "I think so too. Even up to this day it probably is. We had a great group of people. I don't know if it was luck or whatever but we put together one absolutely great and dedicated team."

He was the strength of the program and motivated us to exceed. He was a terrific listener.

Each time that I had a design review with our customer, and we had many of them, he would come to my office and ask me to go over all of my viewgraphs and what I was going to emphasize. He got to know my material well enough that he could have presented and defended it himself. He was knowledgeable in every phase of the program including electronics (he himself was an electrical engineer), mechanical design, optical performance requirements and design, testing, assembly flow, and field operations.

Mike was a keen observer and evaluator of people. As our staffing grew, he created a dynamic and efficient organization. After the program got underway, he had daily stand-up meetings early in the morning in the war room for key people to discuss problems and plans for the day. He asked pointed questions and took no BS. The meetings were very limited in time. People left there with well-defined tasks.

At the peak of the program there were up to 1200 people on the program in many capacities (engineering, manufacturing, test, quality, reliability, finance and planning, administration, etc.). Of course everyone had to undergo the security investigation prior to working on the program. During the time they waited for their clearance, most of them worked in a bullpen area on "sanitized" portions of the job. This bullpen was a noisy place known as the "mushroom patch."

SYSTEMS ENGINEERING

The Systems Engineering department was of prime importance to the program. First of all, they established, in conjunction with the customer, all of the specifications for the design of the whole system. These included the optical, dynamic, structural, electrical controls, thermal, interface, and test requirements. They were involved in all meetings with our customer. They were also involved in negotiations with the customer concerning system performance, schedule, and cost.

Paul Convertito led the department (Fig. 3.2). Paul was superbly qualified, being knowledgeable in the various engineering disciplines. He was also firm, fair, a good leader, an excellent communicator, and a very interesting person. In the old days when smoking was permitted in the workplace, Paul was famous for enjoying a good cigar. Among his staff were numerous talented people.

Paul remembered that Perkin-Elmer could make its own blunders, and sometime they were doozies. The Perkin-Elmer marketing people had an advertisement that was on television and it showed a space view of Earth and talked about all the good science that Perkin-Elmer was doing. There was only one problem: the Earth was rotating in the wrong direction. Convertino remembered the CIA's Roy Burks (Fig. 2.11), who had managed Hexagon's predecessor Corona, kidding him, "Great aerospace

Fig. 3.2 Paul Convertito, head of Systems Engineering.

guys you are! You don't even know which way the Earth is turning!" Convertito remembered someone from Perkin-Elmer's shooting back, "You customer guys are so powerful, why don't you get the Earth to rotate in the other direction?"

Paul Convertito remembered having to take a flight to Washington, D.C. and his boss Mike Weeks said that he would fly down as well, arriving a little later. Paul got to the airport and when he went to pick up a rental car at Hertz they didn't have a car for him. The only thing they had was a bright red Cadillac convertible with leather seats. When he arrived at the airport he saw his boss and honked his horn, but the boss didn't see him so Paul went around the circle again. When Paul came around, he honked his horn and waved, but again the supervisor didn't see him. Finally he got an opening and pulled up to the curb and blew the horn, annoyed that the man had missed him multiple times. The boss came over and was suitably upset by the luxury car. "When I get to see your expense account I'll fix you." But a little later he had changed his tune: "Can I drive it?" he asked.

Another time, Paul was looking at some transmission data with a couple other guys, including someone from the CIA. Perkin-Elmer's award fee was based upon the data. It was suggested that they should create a souvenir to commemorate this particular successful mission. Someone came up with the idea of a currency note. They put Mike Maguire's picture on a Bentledge note, shown in Fig. 3.3 (*Bentledge* meant that it was filled with all kinds of inside information about the program). As Paul tells it,

> So we sat around the table and designed it. It had lots of data information on it and it was green. It said this note is good 96 percent of the time, which was then the reliability performance success requirement for the cameras. People were handing them out at this acceptance meeting. Then the security guys saw it and said "what are you doing?" and they confiscated them, but a few people kept them.

The systems engineers worked very closely with the design engineers in the mechanical, electrical, and testing areas. The Mechanical Engineering department negotiated design changes with them in order to accommodate problems that came up or simpler methods of achieving the desired result. Their analyses also confirmed the structural or dynamic (vibration) ruggedness of designs and made appropriate recommendations.

The teamwork between departments was very good. Naturally there were some disagreements within the company and with the customer, but the great majority of them were solved amicably.

Fig. 3.3 Hexagon project Bentledge note.

MECHANICAL ENGINEERING

Charlie Karatzas was the first manager of the Mechanical Engineering department. He became our hero and role model. All of us thought that Charlie was the best mechanical engineer we had ever known. He was also a great person and a gentleman. He had a wonderful personality and spoke authoritatively about anything. But Charlie had his limits, as project engineer Paul Brickmeier remembers,

> The CIA did occasionally lock some of our people up in a vault at their place as a negotiating tool. I remember one time when Charlie Karatzas, our manager of Mechanical Engineering, was there during negotiations and phoned us in Danbury and said if he didn't agree to something they wouldn't let him out he was going to pee in the corner. And he did!

Eventually Charlie was appointed Director of Manufacturing and Test, and later on Director of Engineering. Len Molaskey succeeded him as department manager of Mechanical Engineering. Len was a fine man and eventually responsible for the design of recovery equipment to salvage one reentry vehicle containing a take-up from the depths of the Pacific Ocean (Chapter 18). Charlie and Bob Jones (Fig. 4.4) switched their roles as Directors of Engineering or Manufacturing and Test several times. Later, Nick DeFillipis succeeded Len Molaskey as Mechanical Engineering department manager.

In most cases the project engineers for each subassembly was a mechanical engineer. The other project engineers were Paul Brickmeier (Fig. 6.1) on the supply, Walter Roman on the platen, Don Fish on the take-up, and me (Phil Pressel) on the optical bar. As the program developed and the staff grew, others were eventually appointed to various subassemblies; Ken Stout was on the film drive; Nick DeFillipis was on the looper and, as mentioned, eventually became department manager; Neil Freeman was on the frame; Joe Prusak (Fig. 7.5) was on the film path; and Howie Cummings was on the pneumatics. Graham Wallace was instrumental in establishing the original configuration of the optical bar and in the procurement of its mirrors. The mirrors were supplied by the Herraeus Company located in Germany. Wallace visited there to make sure that they knew exactly what our requirements were and to set up the proper logistics and inspection procedures for the mirrors.

One member of the Mechanical Engineering department, Jerry Park, was a bearings expert. He was instrumental in choosing the type and sizes of many bearings in the system. He also recommended the proper lubricants and developed a unique method of measuring the running torque of an assembly. Other mechanical engineers over the years were contributors to some complex mechanism designs.

During the early days of the program, Frank Sileo worked on pressure drop analyses to check for system gas leak integrity, system thermal analysis for the nitrogen purge system, and investigated the development of a high-resolution image material for potential system use.

ELECTRICAL ENGINEERING

The Electrical Engineering department manager was Bob Williamson. Dick Carricato (Fig. 2.7), mentioned earlier as part of the Systems Engineering department, was an electrical engineer and made major contributions to Hexagon representing both departments. He was particularly skilled in the design of control systems and derived many of the equations involved in the design of the IMC system.

The electrical engineers were divided into groups that supported the whole program according to electrical function rather than for a particular assembly. The functions included servo control, motors, encoders, packaging, power supplies, cabling, EMI and vendor support for electronic boxes.

One of the most critical electronics was the sequencer box. As this book describes the incredibly complicated system, all of its moving parts, the accuracy of position and velocity that all of the parts had to meet, the reader will begin to appreciate the importance of the electronic controls (Chapter 17) as well as the key mechanical devices. Every step in the control of the mechanisms and the film had to occur at the exact required time and for the proper duration. The sequencer and the servo control (Chapter 11) performed these functions extremely well.

The electrical engineers were also very involved with the Manufacturing, Reliability, and Test departments, as well as with the technicians working in the assembly areas. This included performing actual electrical tests, debugging problems affecting flight or test hardware, and evaluating results of tests and flight data. They were involved in designing and creating state-of-the-art devices without which the program could not have succeeded. Some of these were brushless DC motors, light pipes, light-emitting diodes (LEDs), optical encoders, and methods of servo control.

Among the many technicians working with the electrical engineers were two key contributors. One was Jon Aspinwall, who was a member of the electrical test team and worked on the optical bar servo breadboard, the E-model, the platen servo, the D-model, and chamber testing of the system. The other was Howie Burg, who worked closely with Jon, Marty Yellin (Fig. 11.3), and other electrical engineers. They spent many late hours in getting the camera operational.

For every preliminary and critical design review (PDR and CDR) the electrical engineers (and other departments) supported the mechanical project engineer who was usually the person to lead for that particular assembly.

DESIGN AND DRAFTING

Tom O'Mara was the manager of the Design and Drafting department. Drafting in those days was done on large old-fashioned drafting tables. There was no computer-aided design (CAD)–it was all done with pencil and paper. The drawings were 2-D. There were isometric views, but no 3-D models that could be rotated around like today's CAD drawings.

These were the days of slide rules and pocket calculators. Paper size went from A size (8-1/2 by 11 inch paper) to B size (the double of A size) up to E size (which covered most of a drafting table). Some of these papers were very large, including drawings that were much longer than a table. These drawings primarily showed full-scale assemblies.

Major design changes required much erasing or starting a new drawing from scratch. If a design had to be changed, out came the eraser in the form of a handheld device that rapidly rotated a pencil-sized eraser. For major changes the drawing was completely redone.

There was none of today's solid-state technology such as microprocessors and printers. Copies of the drawings were produced in large machines that contained specially treated paper that was overlapped with the original "transparent" paper and processed using chemicals, primarily ammonia. The lines on the final copies came out in the color blue, thus the name *blueprints*. They were always folded neatly. For some time after producing the blueprints, they smelled of ammonia.

During the design and development of the system, from 1965 into the 1970s, computer technology consisted of large general-purpose mainframe computers such as the CDC 930. It had punch card inputs with print and tape outputs. Dick Carricato performed most of our servo simulations that were used for the proposal (discussed in Chapter 2). His inputs included items like inertia of components, physical constants, and torque disturbances. In order to save time, he created the punch cards himself. Analysis programs were limited to early versions of NASTRAN (a mechanical structural analysis program) and SINDA (a thermal analysis program).

Compared to today's advanced technology, the tools that we used, the methods of testing, visualizing, analyzing, and in some cases in fabrication, were antiques. Most of us engineers used the slide rule and eventually the pocket calculator. Yet we did create an extremely complex and successful system anyway.

Tom O'Mara was a man of great integrity and commitment. He was one of those "yes we can" people that every organization needs in order to succeed. As head of Design and Drafting, Tom was not only a good department manager but he was also technically savvy. He would pick up a pencil and start the design at the moment that he was given the sense of what the product needed to accomplish. He and his talented staff would whip up intelligent and believable concept drawings that would also engender further ideas and concepts. He had the attitude "see what you think of these ideas; if you don't like them, that's okay, we'll keep trying." Tom was instrumental in eventually bringing to our company computerized design and drafting, starting with the design of printed circuits and then into optical and mechanical design. The use of personal computers for design, in great part, came in through Tom's organization.

MANUFACTURING

Kent Meserve headed the Manufacturing department. Kent participated on the program from its beginning. Mike Maguire, our general manager tells this story about Kent.

> We never had any real manufacturing [before Hexagon] except for the optics manufacturing. When we started looking at the complicated assemblies we had to put together here, I started out looking for a manufacturing engineer that had put together fairly sophisticated systems and I couldn't find any. There was a guy named Kenny Meserve, who was a technician at the time, and he had worked at a prior facility down the road. I said to him, "Kenny, I want you to sit down and figure out a flow diagram for the assembly of this whole system." And he looked at me and said, "What's a flow diagram?" I said it is where the assemblies come together part by part until we get to the final assembly. He said "Okay, I can do something like that." He put together a couple of flowcharts and we talked about them and modified them. He was obviously a very bright guy and very intent on doing a good job, and we ended up making him the manufacturing manager for the project. Good guy and a really good decision.

The Manufacturing department engineers were involved in working with the design engineers to make sure that the parts could be made easily and assembled. They were involved in the design of tools to assist in the manufacturing and assembly process. Technicians put the parts together after they were thoroughly cleaned in special ultrasonic baths or hand wiped. The engineers and technicians also performed certain preliminary tests to see that the assembly worked properly.

At various times the Test department was headed by Charlie Bryant, Sven Sathre, and Clovis Prendergast. The department was responsible for producing test plans, creating test procedures, and performing testing for all of the main assemblies on the program. These tests included optical tests, electronic data and control tests, thermal-vacuum tests, vibration tests and any special tests that engineering departments required during the development of the program (see Chapter 14).

QUALITY ASSURANCE, RELIABILITY, PROGRAM OFFICE, AND PURCHASING

The Quality Assurance department was led at various times by Jerry Braddon. The department was responsible for seeing to it that hardware fabricated in-house or at vendors met their drawing or specification requirements. Some members of the department traveled to vendors to provide assurance that the correct procedures and testing were followed.

They also confirmed that the step-by-step in-house testing followed the procedures and test plans.

Stan Karachuk led the Reliability department. It was tasked to obtain failure rates of components (primarily electronic); verifying material properties; performing material tests, such as determining chemical composition of debris in the system; and analyzing bearing lubricants. They also performed life tests on moving parts such as bearings and certain mechanisms.

The postanalysis flight group led by Dan Orlitzki evaluated optical test data taken in the thermal-vacuum chambers and looked at actual film returned from orbit to determine the performance resolution. This data was compared to the requirements of the program. In most cases the on-orbit performance exceeded the requirements. This data was also used, in part, to determine a portion of the company's fees for each flight.

Bernie Malin led the Program Office. This department oversaw the schedule and technical performance of each flight vehicle. It made the decision to proceed after each major milestone. It also participated in negotiations with the customer and associated contractors, as well as communicating special requirements and design changes. A vehicle manager was designated for each Hexagon system and followed its progress from the start to its shipment destined for Lockheed.

The Purchasing department was led by Ralph Sisk. The department ordered parts (mechanical and electrical) from machine shops and vendors. Staff members often traveled to check progress, cost, and schedule. Occasionally the responsible engineer went along to clarify or solve fabrication problems or discrepancies.

SECURITY

The Security department was responsible for making sure that the facility was properly secured and guarded. A special part of the building was kept secure: here the film was processed and the photographs were evaluated for resolution. The Security department's first managers were Frank DeLuca followed by Bob Markin. The Security department's leader for most of the program's duration was Shelley Ferber, assisted by Fred Malmsheimer.

The department initially provided all employees scheduled to work on the program with personal security questionnaires (PSQs). The employee had to fill out the long questionnaires after being hired or prior to being assigned to the program. It was a document with many pages and required you to trace your addresses, schooling, jobs, and overseas travel as far back as possible.

Several personal and business references were required. You were asked whether anyone had ever contacted you to divulge information or offering you something in return for information.

Of course you were fingerprinted and photographed. I eventually heard from neighbors and friends that an agent of the government, usually the Federal Bureau of Investigation (FBI), asked about me with respect to personal habits such as drinking, drugs, argumentative with family or friends, loud parties, etc.

After the employees obtained their clearance (it took between 4 months to a year) they were briefed and had to sign the appropriate security documents and were assigned badges and appropriate badge numbers. Upon entry to the building, an employee had to exchange his company badge for a program badge containing his photograph. When leaving the building, the badges were reexchanged. There was a feature of the program badge number that not everyone had. If one of the numbers on the badge was a zero, that meant that the person wearing it knew who the customer was, namely the CIA. For security reasons not everyone working on the program needed to know this.

Security controlled every single classified document with control numbers and sign-out sheets. All documents after they were issued had to be stamped with the appropriate security tag. "Project Secure" was used only on internal documents never intended for external use. "Secret" was used on most documents that would be distributed outside of our facility. Even notes that we made while visiting another contractor had to be marked secret and transmitted back via courier. "Top Secret" was used for any information that had to do with where and how the system would be used; including identifying any major component and its performance characteristics by its descriptive name, such as optical bar, platen, film, looper, twister, or roller in conjunction with a discussion of a space-based camera. Even referring to the project by its name Hexagon was classified as top secret. One could not reproduce any page of program documents.

As mentioned, couriers were used to deliver classified documents to other facilities. The Security department was in constant contact with government security agencies. For example, they needed to make sure that visitors had the proper up-to-date clearance in order to be granted access to our building. The department also issued rules and regulations on an ongoing basis and provided secure telephones as required.

Sometimes the security precautions seemed like they were taken from a spy novel. Mike Mazaika, systems integration manager and eventual general manager (Fig. 3.4) remembered how he would have to travel to the West Coast carrying classified documents and deliver them to a courier,

Fig. 3.4 Mike Mazaika, systems integration manager and then general manager.

> He told me to meet in the parking lot of a department store. He was the classic FBI guy. Three-piece suit, a tie, and eyes that were absolutely cold. You did not want to get on the wrong side of this guy. Plus, he was big. We met in a different location every time, usually in a busy area. I would take the stuff out of the trunk of my car and transfer it to the trunk of his car. One day there was a glitch. I went into this parking lot and there was nobody there, I mean *nobody*. Here is his car and mine, and we are in the parking lot transferring this material. It turns out that the store was shut down that day because they were preparing to take inventory. He was all charged up looking for cover, but there wasn't any.

Mechanical engineer Mike Krim remembered that there were a lot of airline hijackings in the early 1970s. "If we were carrying classified data, we were told to go to the bathroom and flush everything possible if there was a hijacking," Krim said. Employees received a lot of guidance on where to stay and what flights to take. Krim added, "We were also told where not to stay in Palo Alto, because some of the hotel rooms were bugged."

One time Krim was invited to a dinner by some British friends who wanted him to meet somebody they thought he would be interested in speaking with, somebody who apparently worked for British intelligence:

> So after dinner they brought out cigars, and pipes, and MacCarrans, an 18-year-old whiskey. And this fellow says, "What have you been working on?" I said "I don't think I can talk about it because we really haven't been introduced properly. What have you been working on?" He

> said "I'm sort of in the same situation." So we sat and drank whiskey and didn't say a damn word.

For many years all the work was done in one secure building, and everyone in that building was cleared. Eventually when the building was expanded and other nonclassified programs were being worked on, separate secure compartments were created that were accessed through cipher-locked doors. There were no windows in any working area or compartment of the building.

One got used to the security and "sealed" compartments. There were times when some of us had to work on some "special ultra secret" assignment. These were done either in special cipher-locked areas or in an actual ventilated vault that was the size of a large room. In these rooms or compartments there was the "two-man rule." This meant that a minimum of two people had to be in there at all times. If there were only two people and one of them had to leave such as to go to the bathroom, the other person had to leave and the room locked. To reenter, one of the two people pressed the correct five numbers on the cipher lock or dialed the proper combination of the door lock.

As much as we loved Charlie Karatzas, our director of Engineering, he had one major fault. He habitually forgot to lock his safe at night before leaving for home. Guards checked all the safes after hours. The Security department's general rule was that if a safe was left open, the responsible person got a violation.

After being warned several times, Charlie made himself notes that he kept everywhere, such as on his desk, taped to the safe, attached to his car keys. He still occasionally forgot but I guess due to his position and great value to the company and the program, he was given some leeway. He did eventually get into the habit of locking his safe.

Even though we worked in a secure building and all of us were cleared, we all got used to speaking to each other, whether in person or on the phone, in kind of a coded language. The code primarily consisted of the abbreviations for the major assemblies. The word film was *never* used. Film was referred to as "material." We never referred to any classified words or identified parts with their real names on the telephone.

A typical coded casual discussion between us would sound like this: "The tension sensor outside the FEV is not calibrated, so the material going into the LO is not synched properly" or "there is a once-per-rev torque peak on the OB summed error signal indicating a possible bearing or seal problem," or "the OB is at maximum speed but the material is not following the proper *V/h* signal" or "the BR on the TU says we are almost full" or "there is an offset in the LOS of the A bar in the TCA." There were many variations of coded conversations but they got to be routine for us after a while.

All of our work was covert and we got used to not talking about it with family and friends. We understood the importance of this program to national security. I actually got so used to working in this enclosed environment that I enjoyed it. Sometimes I let my imagination run away a little, and I considered myself a spy because we were designing an ultra secret and powerful high-resolution “spy in the sky” satellite camera.

In the next chapter Mike Krim describes the early days of the program, including our first, failed attempt at designing a camera assembly.

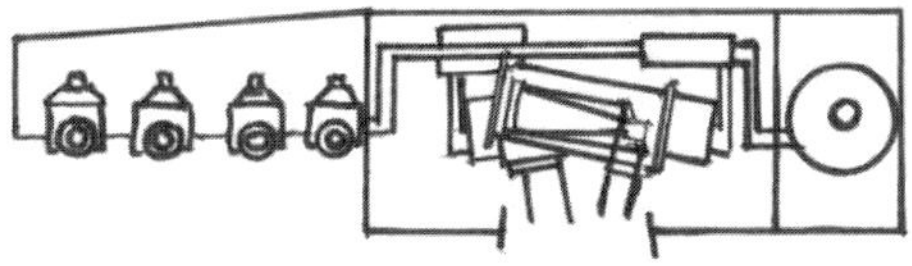

Chapter 4

HEXAGON, THE EARLY DAYS

MIKE KRIM*

We were young once and with the vigor and confidence of that youth, were unafraid of the technical challenges that lay ahead of us (Fig. 4.1). Perhaps this degree of confidence provided some of the spark that led to Hexagon's success. Not that we didn't embark on the wrong roads at times; we did, but were able to get to where we wanted to go although not in as direct a manner as we might have wished. We were entering new territory in terms of size and performance and functional complexity of a reconnaissance camera compared to anything that preceded Hexagon, at least as far as Perkin-Elmer was concerned. Gravity dependence, sometimes a subtle factor in the design of mechanisms and telescopes for ground operation in a 1-*g* field, had to be avoided and a new issue, gravity release, or absence of gravity, in orbit compared to the laboratory manufacturing and test environment caused lots of concern.

There were no handbooks or manuals for the design of large orbital telescopes of the nature of Hexagon so we had to discover and figure things out as we went along. Hexagon was really a mechanism-rich system where a moving image needed to be synchronized with moving film and done in such a way as to reduce "material" (the word we used for film) wastage to just about zero. So in the long run, it was us who actually wrote the design manuals and handbooks! The Hexagon team undertook this task and despite the difficulties it posed, succeeded in a manner exceeding any of our dreams.

Copyright © 2013 by the American Institute of Aeronautics and Astronautics, Inc., All rights reserved.

*The Hexagon program started out innocuously as a study, then a proposal, and finally as a very lucrative contract. For most of us it was a career job. In order to get a better appreciation of those early days, my friend and colleague Mike Krim writes his own inimitable explanation of the system. Mike was a mechanical engineer who came up with innovative ideas on how to configure designs in order to simplify interfaces between various portions of the system, while making sure that they met stringent load and alignment requirements. He was also a fine artist and contributed numerous hand-drawn sketches to help visualize concepts. Sadly, Mike, a wonderful character, passed away in the summer of 2008, before the publication of this book.

Fig. 4.1 Mike Krim, mechanical engineer.

HEXAGON'S SYSTEM ARCHITECTURE

We needed to provide some sort of intermediate storage as the optical systems only needed the film for 120 degrees or less of cross-track scan rotation during photography, separated by 240 degrees or more of "dead" time. The film was delivered from the spools at a constant tangential velocity to avoid starting and stopping the reels—actions that would have been disastrous regarding momentum change and vehicle attitude stabilization. It also would have used up considerable electrical power. Not only did we need this intermittent storage capability, we needed to synchronize the film's velocity vector during the 120 degrees or less period of data (photography) taking. That meant an oscillating focal plane mechanism, referred to as the *platen*. So we had film coming off the supply reels at a constant 70 inches per second that needed to be stored in a buffer, called the *looper*, and pulled out of that device at up to 204 inches per second across the image slit and then returned to the other side of the looper where it would travel to the take-ups spools located in the reentry vehicles at the 70 inches per second velocity. It is worth repeating that the maximum film speed velocity at the focal plane was 204 inches per second. That is equivalent to 17 feet per second, which is incredibly fast.

The exact speed of the film as it crossed the platen rollers had to exactly match the velocity of the image, which was a function of the satellite's velocity, its altitude, the focal length of the telescope, and the angle of the scene from nadir. The film had to be steered over a clever arrangement of air bars called the *twister*. The twister permitted the film to be rotated in plane and pulled across the slit and platen rollers that were rotating with the optical bar. This was akin to "dynamic origami!"

I have tried to convey the idea that Hexagon was a complex, interactive mechanism that had to move thousands of feet of wimpy but wide film precisely past the rotating and moving image produced by the camera without wastage. This was part of the challenge posed by Hexagon. The other challenge was schedule. The program was a very important element in the nation's security and delays in making it operational could carry the gravest consequences.

Motivation and examples were set by management at all levels, and particularly by our leader Mike Maguire (Fig. 2.7). The will to succeed and not let problems defeat us as well as the importance of the program from a national security point of view all contributed to this success. Hexagon was the catalyst that forged a team approach where individual possessiveness (rampant in the technical world) was replaced with positive and constructive interaction and cooperation.

The Cocktail Shaker

We didn't get to this architecture by the most direct route. To use an oft-quoted phrase, "in the beginning" a number of us participated in a program called Ad Hoc that was led by Milt Roseneau and Dick Babish (Fig. 2.1). I was assigned to work with a certain John Rawlings, a Brit in his 50s, on something called the "cocktail shaker" (see Fig. 4.2). That was after a hiatus during which many of us waited and waited, and waited, for clearances and other paperwork to be processed. This was about 1964 and if memory serves me correctly, I had been out at White Sands preparing an Aerobee sounding rocket payload for flight when I received my summons to join the hiatus corps. Getting back to the cocktail shaker, and it did surely did shake, I sometimes reflect on how naive we were.

The purpose of the cocktail shaker was to demonstrate a low-power, low-to-no reaction means of scanning a telescope over a ±35-degree arch in about 1 second...over and over again continuously. It was constructed from aluminum scaffolding tubes, a bicycle wheel, a 6-foot long truncated cone whose moment of inertia (MOl) was representative of the full-scale telescope, and minimal electronic controls (besides an on–off switch). It had no simulated path hardware.

The tubes were two 25-inch diameter and about 100 inches or so long truncated cones representing the optical systems that oscillated +/−45

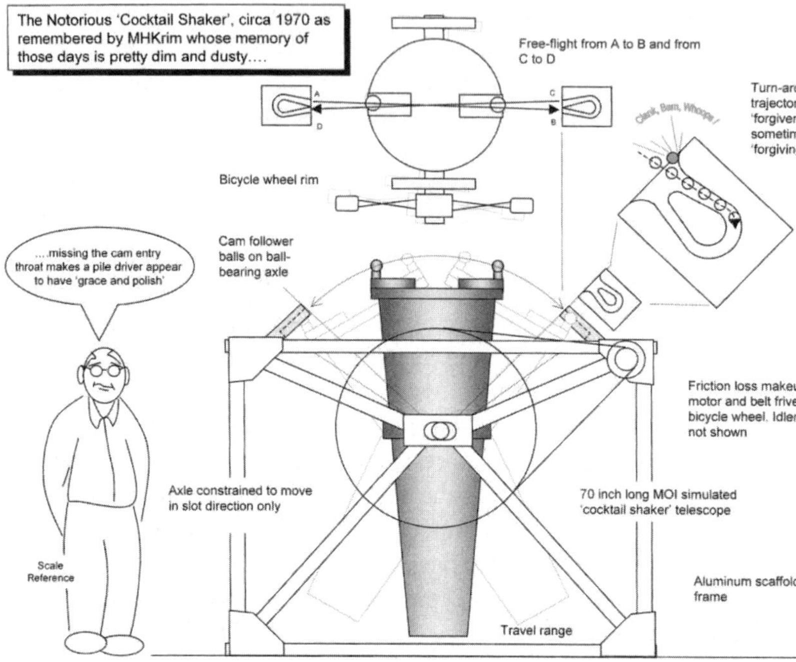

Fig. 4.2 The cocktail shaker.

degrees on axles that defined the 20-degree stereo angle. There were mechanical turnaround tracks at the ends of the swings. These two "cocktail shakers" (hence the name) were synchronized with some sort of a bicycle wheel arrangement but details are foggy. It often missed entering the turnaround track and that resulted in loud banging and clanking noises with John saying something like, "oh dear." He was a Brit and unlike U.S. President Johnson at the time was more reserved in his language.

It was far from a convincing demonstration of a reactionless mechanical oscillator. On the less than infrequent occasions when the cam follower missed the cam entry throat, pile-driver type noises could be heard. John, an exemplar of British understatement, treated these near disasters as "opportunities for design improvements."

The apparatus was mounted in an aluminum framework that was made up from bars and joints used for scaffolding. More creaking and groaning! Maybe it was prescient though as the thing had no film transport capabilities (or concepts for one!) and therefore might be thought of as a forerunner of charged-coupled-device (CCD) focal plane cameras (hah hah). At about the time the thing was banging and clanking away, the Hexagon effort was underway at the Itek Corporation up in Lexington, Massachusetts. What went sour up there I don't know, but the work was transferred, along with test bed hardware, down to Perkin-Elmer in Connecticut.

I can't remember if the customer saw it, and if he did what his reaction was. However it was abandoned shortly after I began working for John for reasons I can only speculate about. I've often remarked that out of partial delusion, rampant as you might recall in the infancy of large space-optics technology, came a number of successful programs. But the cocktail shaker was wisely abandoned before it caused any permanent damage. Despite the misadventure of the cocktail shaker, we did go on to win the development and production contract.

Starting Over

There was a problem with the optical bar's roughly 30-by-20 inch fold flat mirror. It had a 15-by-5 inch slot in its middle, which made it very flexible. Itek had planned to make this mirror out of beryllium (a very stiff and lightweight material) so that its self-weight deflections would be well within the allocation for this effect and thus laboratory testing in 1 *g* would not have been very different from the zero *g* operation. The self-weight deflection was not close to its allocated budget.

We began to look at other ways of configuring the mirror-mounting arrangement so it would meet the allocation. In addition to this 1-*g*/zero-*g* difference, there were other forces present on the mirror due to centrifugal force caused by the rotation of the optical bar. The gravity release issue and the "S"-shaped deflection of the mirror issue resulting from the aforementioned rotation led to the development of the "sandbox" analogy for determining the deflection of mirrors. It enabled the effect of different mounting geometries to be rapidly evaluated experimentally. Today we'd do it with a finite element analysis (FEA) but back in 1965 this kind of modeling was well beyond state of the art. The sandbox led to a mounting concept with four points on the mirror bridged to three points to "ground" and it really seemed like we had a robust and confident solution. Glass would be okay; beryllium, which was very expensive, wasn't needed.

We also didn't like the 3-D truss that Itek had developed to support the camera to the vehicle, although in retrospect it wasn't a bad idea. But we had the idea that the truss, or camera-integrating structure, should be transparent to the film path hardware—in other words be out of the way. That led to the two-camera assembly (TCA) "frame" approach that was interfaced to the Lockheed vehicle structure in a (almost) statically determinate manner.

The film canister containing the two reels of film (one for each camera) was attached separately to the rear of the vehicle, as were the four reentry vehicles mounted just forward of the TCA. So now we had the rudiments of a configuration in hand. We had moved from the "Wilton Hilton" building to a facility across the street that was dedicated to the Hexagon program. That facility had a b-i-g wall. That wall was the "canvas" where I could draw the system full size. This was to orient both ourselves and the new folks, who were coming onboard after being cleared, and to communicate the scale of the endeavor. The "drawing" was actually done with black electrical tape and fat magic markers. Often it was referred to as the "Savior" as most of the people seeing it for the first time would mutter to themselves "Jesus Christ." I say this in my own inimitable style with no disrespect to anyone.

In May 1966 we wrote the proposal for the design and manufacture of the system. I can't remember much in the way of specifics about the proposal described in Chapter 2. I do remember that sometime after it was submitted we were all called into a room where Dick Werner, a small guy with a big cigar who was in charge of the division, informed us that we had won. That win marked the beginning of a lot of intensive and focused work, and team building as well as facility building. It was pretty clear that we would soon need a much larger facility and that it was to be built on Wooster Heights Road in Danbury, Connecticut. Not only would the engineering be done there, it would have one of the world's largest, vibration-free, thermal vacuum facility for optically testing the systems before delivery. Only the optics would be manufactured back in Wilton and then shipped to Danbury for integration. But it would be several years before that facility was completed and ready for us.

THE PROJECT BEGINS

The original contract was for six Hexagons systems to be flown each year and that meant twelve 26-inch diameter primary mirrors and twelve 30-by-20 inch fold flat mirrors would be needed every 12 months. This was in the days of fused silica mirrors. Corning's ultra low expansion (ULE) material hadn't been invented yet. The existing fused silica mirrors had "egg-crate" cores, that is, ribs with notches halfway through their depth so that a core structure could be assembled as illustrated in Fig. 4.3.

It was impractical to "weld" the ribs at the intersections because this sometimes led to buckling when the front and back plates and the core rib assembly were "dry" assembled as a sandwich and heated to around 1800°C to seal (fuse) the faceplates to the ribs. Buckling meant a rejected mirror. At the rate we needed them this was deemed an unacceptable risk. So we looked at a different method of making the mirrors, albeit at a greater weight.

The method we selected was developed by the German company Heraeus and consisted of making a core from a solid disk of fused silica and lightening it by drilling holes that eliminated about 80 percent of its weight. The

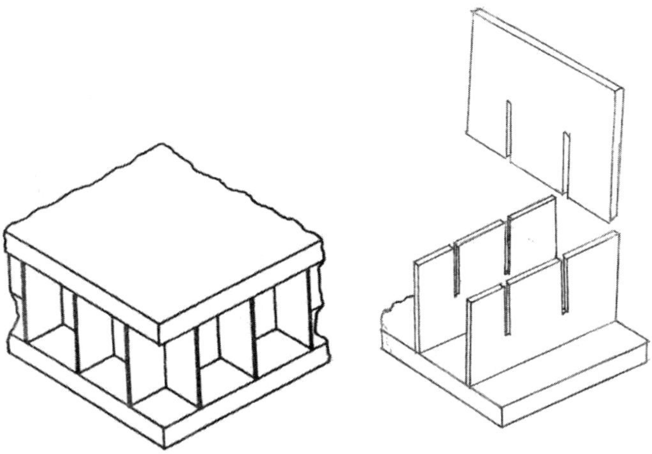

Fig. 4.3 Egg-crate mirror core construction.

faceplates could be confidently fused to this core. I remember to this day asking Graham Wallace, an opto-mechanical engineer, if we could go with this approach and much to my surprise, he agreed. After all, I was a real beginner in this orbital space stuff, having come from Grumman about 5 years earlier where failure was a well-defined event like a wing falling off. Here at Perkin-Elmer, it was a whole different culture. We now dealt with the subtleties of errant microns (a millionth of a meter) or even fractions of microns. In fact nanometers became part of our language so at least we could be dealing with whole numbers!

By this time we were becoming pretty well organized. Paul Convertito (Fig. 3.2), a cigar smoker in those days, headed Systems Engineering and Charlie Karatzas headed Design Engineering. Charlie Bryant headed Test and Verification and Stan Karachuk headed Reliability. All of these people reported to Bob Jones, the Director of Engineering (Fig. 4.4).

The "major domo" over this entire enterprise was Mike Maguire, one of the most technically perceptive persons I have known and a fair and effective motivator. Assisting him was his secretary Addie. I think it was she who really kept the place going!

Before any further work was undertaken, an outside consultant was brought in and giant PERT charts (listing tasks, schedule, and flow logic) were drawn encircling an entire room. These PERT charts and the work that went into preparing them provided a real benefit to the leadership staff. It forced them to talk to each other and to all of the people doing the actual detail design, analyses, testing, and planning. It helped to forge a cohesive team that understood each other's issues. We complained about having to create the charts at the time because it wasn't our primary endeavor

Fig. 4.4 Bob Jones, director of Engineering.

(engineering design or creative analysis), but people like Paul Convertito helped us to realize the value of such an integrated roadmap.

Once the initial program evaluation and review technique (PERT) work was completed, activity resumed on the TCA. We had made our predictions of stiffness and mode shapes with a computer code that we wrote ourselves called MARCK-V. That stood for Martelli, Cochrin, and Krim, with the "V" standing for victory. Models are one thing, but anchor points into reality for the models was another, and that led to the design and fabrication of the "mass" model. This hardware model simulated the structure, its stiffness, and distributed masses of the TCA. It would be vibration tested at General Electric's Valley Forge facility near Philadelphia. We would compare the resulting test data with our predictions. Some wags (wild ass guesses) said we'd be lucky if the "damn units were the same."

This mass model was actually the first frame and optical bar structures that came off the "production line" at the Budd Company in Philadelphia, the vendor who was to fabricate the flight units for these structures. The model consisted of real structures, real bearings, real flexures, real isolators, real universal joints, and real metering bars. The optics and complicated mechanisms such as the looper were surrogates.

As far as comparing the results of the analyses to the test, not only were the units the same, the basic modes were within arm's length of agreement....we were ecstatic.

Another thing that I remember about testing at General Electric was when the laboratory manager, Paul Petty, asked me if there was anything else he could do to assist us with the testing. Being a bit of a snarky young wise guy, I said that cleaning up the lab that seemed to have pools of hydraulic fluid on the floor would be a good idea. I don't think that I would have been so brash had I foreseen him becoming the head of the Hexagon program up in Danbury some years later.

Right after the TCA tests at General Electric in April 1968 we moved up to the new Danbury facility. Windowless from the outside, it was painted yellow on the inside to give the feeling of sunshine. Or was there a sale on yellow paint?

At the same time, and at the suggestion–I mean direction–of the customer, the CIA, that included Les Dirks and two other CIA people, recommended that we bring in two consulting groups to augment our "thin" structural and thermal analysis capabilities. The two consulting companies were Mechanics Research Incorporated and the Aerotherm Corporation.

Next on our agenda was the establishment of interfaces with Lockheed (the company that would integrate our system), the actual reentry vehicles, and many other accessories and electronics into the vehicle that would eventually sit atop the launch rocket. At Lockheed in Sunnyvale, California, we met Bert Bulkin and Dominick Tenerelli and a lot of other experienced people including Ludwig "Dutch" Verugde and Beez Reising, who, despite having hands as large as hams, produced uncannily realistic drawings.

There was a lot we learned from Lockheed; after all, they were the integrating contractor for Corona (the predecessor program) and had real hardware experience. The first integrated mechanical interface drawings were done with me, Dutch, and Beez working in a non-air conditioned office over a weekend. I remember it was hot outside and far from pleasant inside. The supply reels were originally oriented with their axes in the launch direction. We turned them around so that their rotational axes were crosswise in the vehicle. That got rid of some skew bars, and the air that they consumed. It also probably prevented collapse of the 6-foot diameter reels of film stack during launch.

By now our architecture was maturing and approaching reality. We captured the designs and the requirements in the system spec book, another simple but effective way to make sure that all parties back in Danbury had common information. The subsystem designs would be mechanically and functionally compatible. Our company had never been involved with a program of the technical complexity and partitioned multiorganizational responsibilities that characterized Hexagon. I think that it was largely due to the vision of Convertito that the PERT and spec book processes were put in place. Without these navigational aids, progress would have not been as smooth as it was.

It was during this time that new project engineers were assigned to supervise design of the major subassemblies. The film supply assembly was taken over by Paul Brickmeier; the looper and film path mechanical elements by Dick Chester, Joe Prusak, and Nick DeFillippis; the optical bar by Phil Pressel; the frame by Neil Freeman; the platen by Walter Roman; the film drive by Ken Stout; and the take-ups by Don Fish. They were in charge of the design review presentations to our customer, the CIA. The project engineers were all mechanical engineers and were supported by electrical engineers and the other engineering disciplines such as design and drafting, manufacturing, reliability, and quality control. Of course there were many others who helped and made major contributions and eventually took over some of these responsibilities for the next series of vehicle contracts that included improvements.

The film transport control system was Franz Holtzer's and Marty Yellin's domain with significant help from Dick Carricato and Dick Labinger. To aid in the development of the entire film transport system, a full-scale test bed called the film path simulator (FPS) was constructed. It was comprised of a real looper, real film, real reels of film, real air bars and twisters, and a simulated oscillating platen. Gene Gessner had simulated the FPS control functions with an analog computer and Holtzer, with his Viennese accent and propensity for digital techniques, could often be heard screaming at Gene.

Somewhere about this time (1968) we began preparing for design reviews and the preliminary design review (PDR) with the customer. We certainly were in the customer's spotlight (or crosshairs as the case may be) and Vue Graphs were produced ad nauseam and shown with our new backside projection equipment in our new conference room where amusing things tended to happen. A person behind the screen handled the charts and the speaker faced the audience away from the screen that was out of his field of view. This was an area where coordination broke down; especially when a question was asked and the answer from the speaker and the appropriate chart from the behind-the-screen person were out of sync.

After the successful PDR, work continued at a furious pace. I continued working part time on improving the system as well as conceptualizing other cameras. I also was able to finally devote some time to my sports cars (Triumphs) and my banjo.

In the next chapter, colleague Sam Palasciano shares his memories of the program's early days.

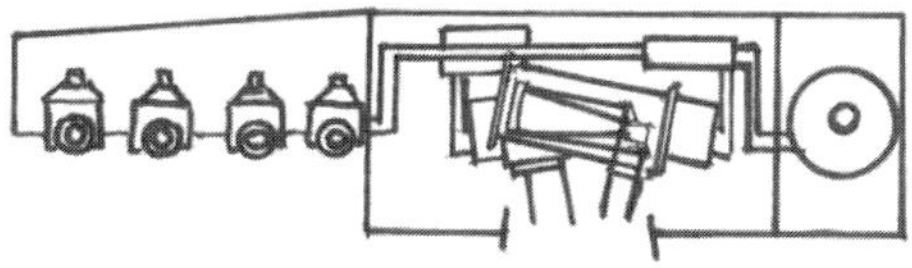

Chapter 5

IN THE BEGINNING AND THE FACILITIES

SAM PALASCIANO*

AD HOC PROJECT

When we first started work on a project in fall of 1965 called Ad Hoc, we were told that Itek was originally planning the program as an outgrowth of the Corona program but they were having difficulty finding a design that would meet all of the requirements. We were selected to compete with Itek for the camera subsystem.

The Ad Hoc proposal team was doing a series of technical trade-off studies evaluating different potential camera configurations and trading one against the other in terms of performance, complexity, costs, size, etc. These studies went on for several months. Periodically, a memo would come out comparing one form of the "F" (Ferris wheel) configuration to another and selecting which configuration appeared best suited to the task. The "M" type system (matchbox) was another camera configuration. Finally in early 1966, a "best" configuration was decided upon. A system spec book was initiated that attempted to define size constraints, mechanical mounting interface, power limits, and thermal considerations, as well as what major components would be required.

In this time period, work was also done to explore electronic packaging concepts. Of course, microcircuits hadn't been invented yet. In the 1960s, state-of-the-art designs utilized high-density, welded cordwood modules. Discrete components and individual transistors were sandwiched between two small circuit boards and connections were welded together. The entire module was then potted, usually with an epoxy-type compound. Subsequent rework was a nightmare. The company had had some experience with small examples of these techniques on another program.

Copyright © 2013 by the American Institute of Aeronautics and Astronautics, Inc., All rights reserved.

*My friend and colleague Sam Palasciano started his employment at Perkin-Elmer at about the same time as I did in 1965. He was first hired to work in the Quality Assurance department. Later he was assigned to work on outfitting some of our test chambers with optical test equipment. He remembers some of the early days on the program and discusses the facility in Danbury, Connecticut.

Dick Carricato estimated that somewhere around 5000 transistors were required for the program. Using an arbitrary four other components per transistor, the electronics were sized at about 20,000 discrete components.

During this phase, we traveled to explore potential electronic subcontractors. We visited Sippican located in Marion, Massachussetts; Spacecraft in Huntsville, Alabama; and Brown Engineering, also in Huntsville. It was almost comical dealing with these companies. They were not cleared for the program so we could not tell them it was a space system, what the nature of the program was, how big it was, etc. I am sure they wondered for our sanity as we left their grounds. Finally, the customer steered us to Harris in Melbourne, Florida. It was then known as Radiation Incorporated. This company was much larger than any of the others and had experience with our customer so our conversations were much more to the point. We selected Harris as our potential subcontractor should our proposal effort prove successful.

One person who was critical in getting Harris to bid on the job was our program leader Mike Maguire (Fig. 2.7). He recalled dealing with Harris:

> With the electronics I had gone to three or four different companies and asked them to bid packaging the electronics, and I could only get one company to bid. That was Radiation, and they kept asking me, "Is this real? We are going to put a lot of effort in this, is it real?" I said, "It's real." And they said, "Who is the customer?" And I said, "No way. You can't talk to the customer; you can't even know who the customer is." So they hemmed and hawed, and I finally got hold of Joe Boyd, the president, and I said, "This is the real thing and you really ought to bid it." And they did bid it, and of course they escalated the price on us quite a bit. I don't know whether that was their fault or our fault, but anyway the price went up substantially. But they delivered a good product.

Sometime in early 1965 Perkin-Elmer purchased a building across the street to house the Ad Hoc proposal team in a secure, contained building dedicated only to that program. As the study advanced and the concepts matured, it appeared that we had an excellent chance to win the job. Mike Maguire, the driving force behind our entire effort, was put in charge of the entire proposal effort. Besides being exceptionally smart, Maguire was an excellent organizer with a very charismatic personality that rallied everyone to work together as he continually kept everything in perspective.

The Ad Hoc team grew in this period and by April 1966 probably numbered about 30 full-time engineers, all reporting to Maguire.

Request for Proposal

The Request for Proposal (RFP) finally came from the customer, the CIA, in early May of 1966. The salient requirements were as follows:

- A stereoscopic panoramic system was needed.
- The system was to have four reentry vehicles.
- The system was to be capable of photographing the entire Sino-Soviet Bloc every mission.
- A minimum resolution of 139 lines per millimeter was to be achieved.
- Each mission was to last at least 39 days.
- The system was to fly in a near-Earth elliptical orbit, 7-day sun synchronous, with typical perigee (closest to the earth) over the Sino-Soviet latitudes at an altitude of approximately 95 nautical miles.
- Film type was to be standard ultra thin based (UTB) black and white film as supplied by Kodak; film thickness would be 0.002 inches thick; and the emulsion was coated on an "Estar" base (Mylar) with a gel coating front and back.
- Scan lengths of 30x, 60x, 90x, and 120x degrees was to be accommodated via ground command; velocity/height ratios of between 0.018 and 0.054 were to be accommodated.
- Six flight systems were ordered plus a nondeliverable engineering model, development model, and crude mass and thermal models.

The Ad Hoc team was now involved at a feverish pace. We had 6 weeks to zero in on a hardware configuration that would satisfy the technical requirements of the RFP, write a description that would convey the idea that our approach was technically superior to whatever Itek was proposing, and develop a convincing cost estimate with back-up to a subtask level. Our future depended on a successful proposal.

Mike Krim (Fig. 4.1) was one of the key contributors in this effort (and in subsequent phases). For one, he developed a configuration that allowed the entire two-camera assembly (TCA) to be mounted on a rock-solid frame. This frame, built like a fortress, was to be suspended on a three-point mount to the spacecraft, so as to be kept coplanar in the event of structural movements by Lockheed's Support System Module. These movements could be induced by either thermal or launch factors.

We proposed a TCA in order to satisfy the stereo requirement. The optical bars (containing the optical elements) were to be mounted such that their pitch angles were each 10 degrees in opposite directions from the line-of-flight axis. This would produce two images of the same target, each taken from two different viewing angles separated by 20 degrees. When set up via a pair of stereo microscopes, the image would appear to the careful viewer in three dimensions wherein the height of the object being viewed was exaggerated due to the 20-degree angle, versus the human eye angle of only a few degrees of separation between the eyes. (As an aside, not all viewers are able to see in stereo due to sometimes subtle physiological conditions of the viewer.) The primary reason for taking pictures in stereo is to provide height measuring capability of objects on the ground.

The film supplies for each camera were to be contained in two 5-foot rolls of live (unexposed) film, 6.6 inches wide. These flangeless (no edge supports to prevent the film from slipping laterally) stacks had to be uncrated, loaded into the supply assembly, and threaded through several rollers—all in the dark so as not to expose the film.

Four recoverable take-up assemblies were to be provided and stored in reentry vehicles. The film would have to be cut upon receipt of an appropriate ground command and reattached to the core of the next take-up. We would be responsible for the design and construction of the film take-ups. McDonnell Douglas in St. Louis would be responsible for the recovery vehicles. These vehicles would be deorbited by the sensor subassembly module (SSM) command and control system and would reenter Earth's atmosphere and be caught by a grapple hook suspended from a four-engine aircraft high above the Pacific Ocean north of Hawaii.

These recoveries were to prove remarkably successful over the program's 15-year history with only one failure, which I believe was due to a malfunctioning parachute. This leads to a whole other story discussed in Chapter 18.

A vacuum test was necessary for final acceptance of the TCA. This required a vacuum chamber (all stainless steel) 50 feet in diameter. Smaller vacuum chambers would be needed to test the optics as an assembly, to test the optical elements individually, and to do some developmental tests. All in all a collection of five state-of-the-art vacuum chambers would be required for the program.

A simulation of launch mechanical and acoustical vibrations were required, leading to eventual construction of an acoustic chamber and random vibration facilities capable of producing some 30,000 pound of thrust.

The government required that the company provide the land as well as the bricks and mortar necessary to support these facilities and they, the customer, would pay for all of the special equipment needed to support the program.

Our proposal was submitted around the end of May 1966. It amounted to many millions of dollars (a cost plus incentive fee contract) and an eventual staff of over 1000 people. The program was to be completed in about 3-1/2 years, from start to finish.

Keep in mind that the entire corporation was probably doing about $50,000,000 per year then and we had about 75 technical people available to the group. We needed to hire almost 1000 engineers within a year—sooner if possible as clearance time was enormous.

Customer Review of the Proposal

Around July 1966 the customer, the CIA, announced to us that they would send a team of people to us to review our proposal. A 2-day meeting was scheduled. The customer, consisting of agency experts, Air Force experts, and people from the TRW Corporation (the Systems Engineering Technical Support contractor known as SETS) came in for a detailed presentation by us to be followed by one-on-one discussions with all key personnel. The CIA technical team was led by Les Dirks, a high-level CIA manager with a reputation as a very smart engineer/physicist. Also in attendance from the CIA was John Crowley, who was the Corona program manager and then became the first chief of the new CIA Office of Special Projects. Don Patterson, who became the CIA's program manager for Hexagon, was also in attendance. His close association with Mike Maguire and willingness to compromise helped a lot in making the program work so well.

After several dry runs, the last being on a Sunday before the big review, the proposal review meeting with the customer began on a Monday. The meeting was kicked-off by our CEO Chester Nimitz Jr. giving a talk about our commitment. He sat up at the front with his feet on the desk and discussed the cash commitment we had committed up until then. We had already purchased the site across the street and additionally had procured a 26-acre parcel of raw land just 15 miles to the north. We were prepared to break ground the day after we were awarded the contract. We had already contracted with Turner Engineering, a major design/build corporation to provide us a turnkey facility of some 250,000 square feet in about 12 months. Pittsburgh-Des Moines was lined up for the vacuum facilities to be installed during building construction.

As to the 1000 more technical people needed, Chester, using flowery four letter words throughout his talk said that we would get them and ended his talk with "we want this f- - -ing job." With that talk as a lead-in, our quaking team began to unfold our technical proposal. Each aspect of our proposal was presented and subjected to intense scrutiny by the 75 or so government experts in attendance.

My recollection was that the technical presentation went well. Near the end, Mike Maguire presented the organization chart. A typical matrix organization was shown wherein the program managers were key to running the program. Bob Kelley, Maguire's administrator during the proposal, came under extensive attack as he was shown to be the manager of the group of key people who wrote the proposal and were scheduled to work on the program. Kelley was under attack because he was not an engineer and the CIA questioned his technical depth. After the next break Maguire came back with a revision replacing Bob Kelley with Wally Benson. Our people were taken aback by this as we weren't sure Kelley wasn't better suited to the job and Benson was one of the few remaining senior managers left in the Electro-Optical Division, which still had a plateful of meaningful work to be done. This personnel change clearly showed the customer we were prepared to pull out all the stops to meet the program needs. Kelley remained on the program until the contract was issued, and he was then transferred to another program.

The second day consisted of each discipline assigned a different room and the customer team members free to walk into any room and discuss that segment of the proposal. A member of the NRO staff, Colonel Frank Buzard, also visited us. Buzard asked "who was going to stand up just before launch and swear the system was launch-ready?" The answer came back that it would be the program general manager or his representative.

The Project is Awarded

On October 10, 1966 we were all called to a general meeting at 10 a.m. in the drafting/engineering room. By this time Dick Werner, Nimitz's former executive officer on WWII submarines and also an Annapolis graduate, was named Ad Hoc general manager. Mike Maguire was head of operations. About 30 people assembled at the appointed time. When we were all assembled, Werner strode in with Maguire and took out a huge glass tube with a cigar inside. He deliberately opened the tube, took out the cigar, lit it, and after a big exhale, announced that we had won. The room erupted in pandemonium. Everyone raced for a phone to call their spouse or whoever; the smart ones called their stockbrokers. Our stock was to split seven times in the next few years, each time climbing above its last price. We were to be off on the technical adventure of a lifetime.

Housing Hexagon

During the proposal phase, the Hexagon personnel were housed in a temporary facility across the street from the main building of the Electro-Optical Division in Wilton, Connecticut. We had purchased a large site 15 miles to the north in Danbury in the event we were to become the successful bidder.

The Building

Upon notification of contract award, plans were immediately solidified to build a suitable plant for the design, assembly, and test of the Hexagon system on that site. The assembly and test plans determined the necessary facility requirements.

Turner Construction, a notable large-scale design/build contractor, was awarded the building design and construction contract. I believe they were given incentives to meet a very stringent schedule. The total square footage involved was about 240,000. This included about 20,000 square feet of a class 10,000 low-bay clean room for detailed subassembly fabrication and 50,000 square feet of a class 100,000 high-bay test area. Class 10,000 meant that the area had to contain less than 10,000 particles, 0.5 microns or larger per cubic feet of air space. Similarly, class 100,000 meant the same for 100,000 particles per cubic feet. In aerospace jargon, a "high-bay" is a room with a high ceiling and usually an overhead crane for moving large pieces of equipment.

Mike Maguire shares the following story about the facility:

> It was to be a classified facility and the town people had to get permits and so forth. The tax assessor evaluated the value of the facility at $10 million. And he then said it was his experience that the equipment in a big factory like that would be equal to the factory itself. So he wanted to assess us another $10 million. I said, "I'm sorry but most of the equipment going in here will be government-furnished equipment and you can't tax that." He didn't believe me, and I got a call from Senator Lowell Weicker who was the U.S. senator from Connecticut at the time. And he said, "Would you come down and see me as you are having some difficulty from the people in the city of Danbury." So I said sure, and I went down. And he said, "What are you building up there?" And I said, "I'm sorry I can't tell you what we are building up there." And he said, "Why not?" I said, "Because you are not cleared." And he said, "I'm a senator, I'm cleared for everything." I said, "Look, I'm sorry, I don't want to argue with you. I can't tell you the stuff, but if you need it call the Department of Defense and they will put you in touch with the right people." I never heard anything more about it and we saved that money.

Thermal-vacuum Chambers

The high-bay test area was characterized by two large thermal-vacuum chambers and associated test equipment. The Pittsburg-Des Moines Company designed and built the chambers onsite. The major chamber (chamber A) was a 50-foot diameter sphere made entirely of stainless steel. It was one of the world's largest spherical thermal-vacuum chambers. Final test of the TCA was to be performed in this chamber. The test configuration included the TCA and supply assemblies mounted into the vehicle midsection, supplied by Lockheed Missile and Space Corporation. A simulated take-up was mounted in an airlock so film could be processed without breaking the vacuum configuration.

Six collimators (optical instruments) were mounted in the chamber, three for each camera, so that as an optical bar rotated, it viewed each of the three collimators at three different field angles. The collimators were Cassegrain telescopes with apertures of 22 inches and a 420-inch focal length. Each collimator contained an elaborate reticle that was projected into the camera system and was used to assess the optical quality of the lens–film combination. Image smear as well as resolving power was measured by flashing first one part of the reticle and then the second part. If the film and aerial image were synched together, no spatial errors would be present.

Both chambers were further unique in the arrangement of the supporting structure that the camera system attached to. Analyses showed that disturbances coming from the ground, such as trucks passing on the nearby highway, needed to be minimized in order to accurately measure the camera performance. This was accomplished by locating the center of gravity of the TCA support structure such that it was coplanar with the mounting plane of the TCA. Therefore, a large ballast cylinder was hung from the support structure recessed below the ground–structure interface. This cylinder was filled with hundreds of thousands of pounds of shot.

The second large chamber (chamber B) was a 30-foot diameter, 50-foot tall stainless steel thermal-vacuum chamber. It was intended that the TCA alone would be tested in it. Early in the program this test was deleted as being unnecessary and this chamber was not utilized for this purpose. It was eventually used to test the Hubble Space Telescope primary mirror years later.

Chamber A was cryo-pumped. Mechanical roughing pumps brought ambient pressure down to the 0.01 torr range (1 torr is equivalent to 1-millimeter mercury or 1/760th of an atmosphere). Then the thermal shrouds, cooled by liquid nitrogen, condensed out the remaining air molecules so that orbital vacuum pressures of approximately 0.000001 torr could be achieved.

CONTROL ROOM

Test cables from the test chambers to a nearby control room were routed beneath the floor. The TCA subsystems were controlled by a hardwired computer-based test system known as the SSTC. This machine was designed and built by the A-C Spark Plug Company in Racine, Wisconsin. Analog recorders were utilized to record the 156- kilobyte data stream. Post-test processing converted this tape to digital format to allow off-line analysis.

Low-bay Area

The low-bay area consisted primarily of class 100 hooded clean-room benches and worktables. The benches were used for the assembly of very delicate and contamination-sensitive hardware such as bearings, mechanisms that went into the platen and film drive, motors, encoders, and some electronics. The low-bay area also had a darkroom. This was used for assembling the film rolls onto a supply assembly.

This entire elaborate 240,000 square foot operational facility was opened for occupation in the spring of 1968, about 2 years after the initial proposal was submitted. It would start assembling and testing flight hardware in the near future and continued to do so successfully for approximately 18 years until the end of the program.

The building was a hubbub of activity. The following chapters describe in detail the camera parts designed for the Hexagon program, starting with the supply assembly in Chapter 6.

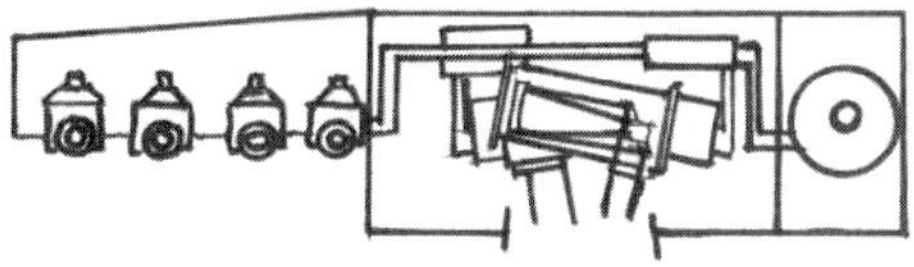

Chapter 6

Film Supply Assembly

Paul Brickmeier and Phil Pressel

The project engineer for and chief designer of the film supply assembly was Paul Brickmeier (Fig. 6.1). In addition, he was involved in negotiations and frequent interface meetings with the CIA and the supplier of the film, Kodak. He remembers the beginnings of the assembly design concept by telling a story about John Rawlings and the cocktail shaker described in Chapter 4.

> John was a Brit and a tinkerer, the kind of guy that made working pendulum clocks out of cardboard. One day he was given the job of designing a model of the system for taking film up to the reentry vehicles. It was far more complex than what they were actually designing, and for that reason Rawlings loved it. He had a small army of people working on it whereas I, who was designing the baseline design—the one they actually planned on using—only had one other person to help me. So I went to Maguire and told him he needed to stop Rawlings, because his own team was getting discouraged. Maguire told him that Rawlings design was never going to get built. Sometime later, Maguire got Rawlings' project shut down, and instead gave him the job of working on cutting and splicing film.
>
> I was concerned that John would be upset that his project was terminated. I thought John would be down, but he wasn't upset, he was happy as a clam. He said "Wow! We're gonna cut and splice!"
>
> I remember that John later designed and built a motorized scale model of the near-final camera system design. He even put in a coin machine so that people had to put a quarter in to see it operate and he even made Chester Nimitz put money in it.
>
> But Rawlings' inventiveness had its limits. He had a Chinese junk—a type of boat—and one day he invited several of his coworkers to go out to sea with him. They arrived at the marina, but the junk was gone. All

Copyright © 2013 by the American Institute of Aeronautics and Astronautics, Inc., All rights reserved.

Fig. 6.1 Paul Brickmeier.

they saw was the top of the mast sticking up from the water. And John, as a master of understatement, said, "Oh my."

At the start of the program (after the contract was awarded to us on October 10, 1966), challenges extended to matters beyond technology. The need to achieve a staffing level of over 1000 scientists, engineers, program managers, and skilled technicians was of the highest priority. At program inception, the entire staff of the program-assigned personnel was engaged in the recruiting process. It was not uncommon for engineers with a few years professional experience to interview candidates for senior management positions. The recruiting process was coordinated by one senior HR manager and proved so successful that it was the major factor in bringing the assigned staff to approximately 1200 people within 15 months.

PLANNING THE HEXAGON PROJECT

The company had no experience executing a program of the scope and complexity of Hexagon. Initial attempts at program planning were, as anticipated, inadequate and yielded overly optimistic predictions of program completion and resource needs. Given the organization's immature planning, the CIA and our program management staff engaged in an expedited, but deliberate, process of insisting that program and task management engage in a process of planning education, plan generation, review, and regeneration. That process lasted for six months or so. A planning "war room" was established and the planning-refinement process was so intense that planners were walking on a few inches of paper as the hand-drawn program evaluation and review technique (PERT) charts were torn down and replaced at a furious pace. In the mid-1960s computerized planning/scheduling tools were in their infancy, being mainframe-based; therefore, they tended to impede the frantic pace of plan development. Upon plan approval, computerized tools were used to keep control and facilitate development of needed workarounds.

Paul Convertito, manager of the Systems department, insisted on us creating good plans and PERT charts. Planning using PERT identifies in chart form specific activities and milestones, the sequence of the activities, the time required, and the critical path. The latter is the longest path in the project. The chart were be updated as progress was made.

Contractors Complicate the Picture

Some felt that the process was unwarranted and that the planning effort was absorbing resources that needed to be applied to the design and development effort. Those concerns were heightened by the fact that the planning process was executed in an environment that varied from collaborative to combative. The CIA had engaged the TRW Corporation, known henceforth as Systems Engineering Technical Support (SETS), as its supporting oversight contractor. In the first phase of the program SETS management, under the leadership of Bill Besserer, was aggressive in its attempts to find the sensor contractor deficient in its program execution. The SETS organization was motivated by a desire to posture TRW as the obvious alternative for management and execution of all or part of the Hexagon effort.

Another contractor with a large role was Lockheed, the system integrator. Given the modular subsystem architecture of the sensor system, had the integration taken place at Lockheed, our role would likely have been limited to subsystem design and manufacture with Lockheed claiming control of sensor system design responsibility.

Those challenges notwithstanding, Perkin-Elmer managed to maintain its position as the sensor-integrating contractor and to persuade the government that the sensor-to-spacecraft integration take place at our facilities. The latter was key to maintaining control of the sensor system integration and test activities as well as the company's maintenance of its position as an "associate contractor." An associate contractor has a direct relationship to the government contracting office as opposed to a subcontractor designation wherein the relationship is indirect. Perkin-Elmer might have ended up a subcontractor if Lockheed had garnered the sensor integration role.

Early in the program, the customer and contractors assigned to oversee our efforts outnumbered our company's program staff by three-to-one or more. Subsystem lead engineers were faced with fielding a barrage of planning and technical questions while trying to make development progress.

Given the covert nature of the Hexagon program, associate-contractor, customer, and customer-support contractor identifications were subject to need-to-know rules. Secure communication requirements were enforced. Initially this meant that all classified telephone discussions took place via a limited number of secure lines that were managed by specific personnel. Of necessity, these conversations employed numerous acronyms that at first were difficult to cope with. A breach of the classified communication protocols would result in immediate breaking of the line, (hanging up the phone). As a result, most productive communications were those conducted in face-to-face meetings held primarily at our facility. Any new outside personnel involved in face-to-face discussions had to be "introduced" to us by a third-party person known to have the proper security clearance and who could vouch for the new person.

The location for these meetings at our facility was driven by the need to maximize the productivity of our available staff, so all concerned were encouraged to conduct business in this manner. Encouragement took the form of surprisingly steadfast and effective insistence at the highest levels of our management and so the CIA, TRW, the Aerospace Corporation (another active and valuable think tank consultant nicknamed Circle A based on its logo), and Lockheed convened required meetings in our building. Eventually we had field offices at Lockheed in Sunnyvale, California and at some other sites.

Perkin-Elmer Staff

The meeting schedule that oversight spawned was so pervasive that extraordinary effort was required to assure both internal technical review and appropriate coordination within the company. It was common for the key staff (typically 10 to 20 percent of personnel) to extend their scheduled 60-hour workweek by 10 to 20 hours per week. Much of that time was spent conducting internal reviews and meetings after hours when the CIA and oversight folks were not onsite. Meetings with program management often convened at 7:00 p.m., with sandwiches brought in for dinner at about 9:00 p.m. while the meeting was in progress, followed by adjournment at 11:00 p.m. The meetings became known as "Addie meetings" because it was the program director's secretary (nicknamed Addie) that would be tasked to call in an order at a local delicatessen. The meeting environment was challenging but collegial.

The program director and division general manager, Mike Maguire (Fig. 2.7), was a quick study and could bore into technical and program issues with an uncanny ability to get to the heart of the matter efficiently. In addition, Mike seemed to be graced with total recall and those that tried to get one past him were likely to be admonished with a "listen, buddy…" and a reminder that at some previous time they had made a different commitment. Mike's work ethic was tough to duplicate; he was typically the first one in and the last one out of the plant.

Company efforts to meet the challenge of executing a program of Hexagon's magnitude resulted in a number of missteps, but CIA oversight was successful more than once at redirecting efforts as required. At one point early in the program the CIA's John Crowley, chief of the Office of Special Projects, and Leslie Dirks made an unannounced visit to our facility asking to conduct one-on-one interviews with the platen and film supply unit subsystem task managers, a few department managers, and division general management. Les Dirks, a high-level CIA staffer, was well known as a dispassionate and hard-to-read individual, and the interviews he conducted were consistent with that reputation. Questions ranged from seeking each task manager's understanding of assigned tasks to identification of the names of our personnel that were perceived by the interviewee to be key to successful completion of those tasks.

At the end of the day, John Crowley and Les Dirks as well as the film supply and the platen task managers identified the key people of different disciplines to division management. This emphasis on key people became an important factor in the relationship between Perkin-Elmer and the CIA.

Kodak Supplies the Film

Film is discussed extensively in this book. Due to the great secrecy of the program, using the word film would obviously mean that a camera was the subject of the program. The word film was therefore taboo and never uttered. Instead it was always referred to as the *material*. The film was 0.002-inches thick (later in the program, thinner films were developed by Kodak thereby permitting more film to be used) and 6.6-inches wide. The film had no sprocket holes. It was driven forward or backward by the film drive motor with the whole width of the film in contact with the motor drum.

The film supplier (Kodak) interface proved to be unique in that interactions were limited to a select few CIA, Perkin-Elmer, and Kodak personnel. Initially, interface meetings between our company and Kodak were actively managed by the CIA; no contractor meetings were allowed without their participation and leadership. Historically Kodak had been independent to the point where they had never contractually agreed to a comprehensive film/film packaging specification of the sort desired by us and the government.

Early in the program, one of the CIA managers led the interface with Kodak with one principal support person. On Kodak's side the key manager was an ex-Air Force officer named "Big" Ed Greene, who was in turn supported by a crusty and shrewd customer interface manager. Early meetings, held at Kodak's facility, consisted of tests of will between the CIA and Kodak regarding who would control the meetings and their agenda. From our perspective, it was clear that there was lots of baggage brought to the interface by both sides but the specifics of that history were never revealed. Needless to say, those early meetings weren't very productive in terms of establishing a suitable specification for the film that was wound on a core (like a spool) designed by our company.

At the end of one particularly contentious meeting, when the CIA manager started to summarize the meeting outcome (so that he and others with him could make the next flight back to Virginia), Kodak's Big Ed decided that a caucus amongst Kodak personnel was required immediately. He proceeded to withdraw all Kodak personnel to a meeting across the hall. With the door wide open they had coffee and seemed to be killing time rather than doing anything of great import. Understandably, this enraged the CIA manager. After 10 to 15 minutes of this the CIA manager announced that he and his entourage were leaving "this monopoly," and they proceeded to do just that. In a fortuitous move, Paul, as the supply unit's project engineer for our company, requested approval to continue the meeting without the CIA's presence, and the CIA manager begrudgingly agreed as he left the conference room. As soon as CIA folks cleared the room, Big Ed returned with his entourage and the information exchange between the Kodak and our people turned instantly productive. Thereafter we held meetings without CIA participation. The CIA subsequently assigned a representative for working-level meetings who seemed to carry no baggage, and progress toward a film/film packaging specification was steady and deliberate.

That being said, Kodak engaged in the film specification activity as a means for reaching an understanding as to the program needs, but made it very clear that they had no intention of ever signing up to it as a contractually binding specification. They claimed never to have signed up to other than a best-effort sort of agreement.

The early conflicts surfaced one more time, when to everyone's surprise Kodak hosted a large dinner meeting and at that meeting, further surprising all concerned, by agreeing to a contractually binding film/film packaging specification—with one proviso. That proviso being that the CIA manager's support person (inexperienced and unnamed in this book) would never darken their door again. A bit crass, but nonetheless the CIA agreed and the deal was struck!

Like Perkin-Elmer, Kodak also had a top secret facility. Paul recalls an adventure while following security instructions when visiting Kodak.

> I remember the time I had to go to Kodak in Rochester, New York. We were trying to move towards not having a caging mechanism since the film was unlikely to telescope on itself. Kodak had done some testing and I was told to go there, but to remove all identification in my hotel room. I didn't have a driver's license, no identification of the company, nothing; I drove the rental car to a designated location and parked and waited. A Ford pulled in front of me with a license plate number I was told to watch for. I followed the Ford, and we go along this big industrial complex with a fence 15-feet high, and the Ford slows down and I slow down behind him. All of a sudden this guy comes out of a door and unlocks the gate, opens it, and the Ford pulls in and I pull in. A guy comes out of the Ford. I don't know who he is—no introduction, he gives me a badge for the company with my photo on it. They take me in through the door and they tell me to stand in the hallway. They disappear. I'm standing in the hallway alone. Now I'm thinking "How am I going to explain if somebody comes around? You don't work here. Where did you get that badge? What's your name?'" But before he waited too long a man showed up. The guy gave him the information necessary for the design, turned around, and left. I stood in the hallway and the security guy comes and unlocks the door; I got back in the car and I left. That was it. I was incredulous but it turned out to be not the only time we went with no identification at all.

Bob Kohler, a CIA manager, remembered that their relationship with Kodak had its lighter moments.

> We built a whole facility up in Rochester for the postflight analysis activity. It was a nice facility, a semi-clean room with light tables, a microscope, resolution targets, and a conference room in the right-hand side where the group could meet. We had an office that was supposedly my office, and I told Kodak (they were really cheap in those days) that I expected this office to be outfitted with vice president-level furniture—-mahogany desk and conference table. When we got there for the very first postflight analysis meeting [to analyze photos taken by Hexagon], they had an unveiling of this office and they had somehow found an old icebox [as well as] the rattiest desk and conference table you could find and they had set them up in the office with a picture of George Eastman above the desk. "Here is your office," they said. They replaced it with good stuff later. I don't know where they got the old icebox.

Another unexpected turn of events took place during one meeting in the early 1970s at Kodak. This was at the beginning of the contract for the second set of six systems. During the middle of a meeting an announcement was received that the program had been transferred from the CIA to the Air Force. This was being done because the program by then was operational and the launches were being run by the Air Force. It also enabled the CIA to concentrate efforts on newer reconnaissance projects. The transfer proceeded very rapidly with a minimum of problems.

The impact of the customer transfer from one government agency to another on the Hexagon program was significant. The Air Force approach was risk averse to the point where even low-risk improvements proved difficult to make. In contrast, the CIA was prepared to take risk in proportion to the apparent benefit. The Air Force 2-year rotational cycle for personnel seemed to instill a culture of "no change on my watch" as opposed to the CIA, wherein assignments were longer lasting, personnel were more accepting of risk-taking assignments, and individual performance assessment was more clearly linked to achieving mission performance objectives. The one significant change that took place after the transfer was the incorporation of a star-tracking device on the camera support frame of the sensor subsystem. The star tracker, called S-cubed (a spaceborne stellar sensor), designed by our company, provided the attitude reference needed to utilize the wide coverage for map-making sufficiently accurate to meet the needs of future technologies such as smart bombs and cruise missiles (see Appendix E). The original systems had a film star tracker made by Itek.

Configuration

The supply assembly was originally configured with its material roll rotational axis coincident with the vehicle launch axis. This was changed later at a meeting between Perkin-Elmer and Lockheed. It was at that meeting where agreement was reached: the film subsystem would be located aft of the camera, with its axis perpendicular to and intersecting with the vehicles longitudinal axis (see Figs. 6.2 and 6.3).

Mike Krim was very influential in promoting this decision. In this configuration with the two rolls of film rotating in opposite directions, unbalanced momentum disturbances associated with angular acceleration of two half-ton film rolls are almost cancelled out and any remaining disturbances are imposed on the least-sensitive vehicle axis (pitch). Another advantage would be that the film path from supply towards the camera would be less complicated and more direct. So it was advantageous to both our company and Lockheed. As Mike remembered,

> It was Lockheed's idea to rotate the supply reels. This was not done as part of negotiations. It was on a weekend on a hot sticky day in a un-air conditioned room at Lockheed with Dutch Vrugda and his designer named Rising. I went out there with the idea that the supply assembly axis should be perpendicular to the flight direction with snubbers to prevent telescoping of the film layers. Telescoping of the film meant that inner and outer layers of the film roll could slide relative to each other and create a nonuniform shape of the film stack. (It turned out later that telescoping of the film never occurred.) That design looked pretty good. It really had to be that way otherwise momentum disturbances would have meant that Lockheed would have had to provide a huge amount of motor power to cancel out these disturbances that would have pitched the whole vehicle back and forth. So not only would the supply reels rotate in opposite directions as we had decided earlier to rotate the optical bars and the take-up reels in opposite directions to compensate for momentum effects.

This configuration allowed the fully loaded film supply unit subsystem to be "assembled" into, or removed from, the spacecraft through a large structural "door" in the aft section of the spacecraft structure as shown in Figs. 6.2 and 6.3.

Film Rolls

Initially, we hoped that it would be possible to space the rolls so that film transport would be straight through to the take-up reels at their lateral offset (having the same center-to-center distance between the supply reels, take-up reels, and the cameras or optical bars). This proved impossible due to two reasons: 1) the camera sizes and spacing were wider than the supply and take-up reels; and 2) the size of the structure necessary to support the large quantity of film through the launch environment, while maintaining film roll

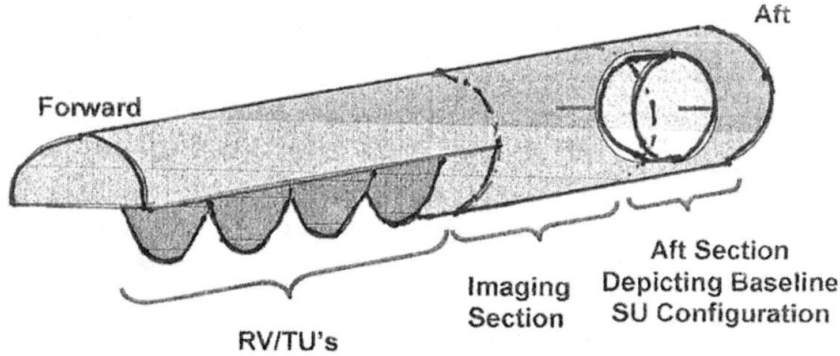

Fig. 6.2 Supply carries film and transfers it sequentially to each take-up.

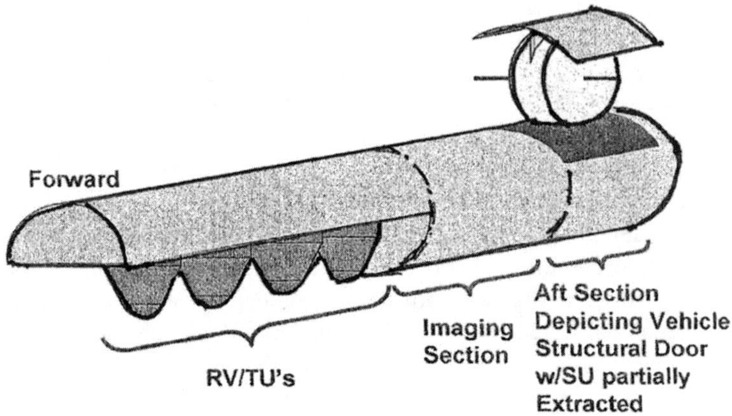

Fig. 6.3 Supply baseline configuration depicting structural door with film supply container partially extracted.

integrity and providing a configuration that supported the ability to load the supply assembly in a darkroom environment, was quite large.

The film was loaded in virtually total darkness. A low number of unique, low luminescent reference lights were spaced around the darkroom to provide a spatial reference for those working in the room. This lighting was so low that even after an hour of adaptation the only thing that could be seen was the shadow of object directly in the line of sight to the luminescent surface. Given clearances on the order of 1/8 inch proved to be easier than initially anticipated. In fact, sophisticated slings attached to overhead cranes and manual guidance replaced automated solutions that we initially proposed. Over the course of the program, the installation and removal process resulted in no damage to the film rolls or supply hardware. Film-load change-out from a fully integrated spacecraft/sensor system was typically accomplished within 24 hours, including its recertification for flight worthiness.

Early in the Hexagon program's development effort, a test film roll was exposed to a vacuum environment and was observed to telescope spontaneously. The cause for that behavior was rapid off-gassing of water entrapped in the film's emulsion layer reducing that layer's thickness and creating an air-bearing effect between layers, which in combination allowed small residual forces to move the wound film web laterally. It was subsequently concluded that the film roll must be placed in an environment that maintained a water-vapor partial pressure consistent with the as-manufactured relative humidity of the film roll. In other words, the film roll would act a source for water vapor and the supply would be passively pressurized. In turn, the supply subsystem had to provide a sealed enclosure

Fig. 6.4 Centrifuge test of boxed supply assembly with the test article mounted to the boom.

that could support a pressure differential of not less than the partial pressure of water vapor at 70°F. Further, the supply needed to restrict the lost water vapor to the downstream film path: therefore, the entire film path had to be sealed.

Film roll structural concerns included the possibility of launch-induced film center-of-gravity shifts. In response to this concern, our company conducted a centrifuge test with the largest film roll that could be wound at the time. Film was wound onto an aluminum core at the highest available winding pressures and subsequently installed in a structure simulating the proposed supply assembly structure. That assembly was in turn enclosed in an aluminum jig plate box for security reasons. The test was conducted at MIT Lincoln Laboratory's centrifuge facility at Hanscom field in Massachusetts (Fig. 6.4).

The CIA wanted to see if the film stack would telescope or slide out of its disk shape. So we at Perkin-Elmer put the film supply stack into a box and took it to MIT's Lincoln Laboratory facility. The CIA had cleared everybody at the lab, but we thought that the MIT staff were kind of professorial and not entirely to be trusted to keep completely quiet about the test.

The CIA provided security standing outside, and there was some concern that if the test article came loose, it might go through the wall. You don't think much of that until the centrifuge gets wound up and comes past the window and you wanted to duck. A thousand pounds of film on the swing arm would pack a lot of momentum if it came loose.

Paul Brickmeier was present for the test and tells the story.

> The guys in the lab were absolutely pissed off that they didn't know what was in the box being tested. I think that if we had driven a Volkswagen in from the street and mounted it to the centrifuge they would have done it in a half hour. They knew what was up though. They had probably done work with Itek before, but the fact that we wouldn't tell them had them slightly miffed, so they did everything reluctantly.
>
> The test went fine; the film, tightly packed in its spool, did not telescope. I remember the worst part of it was when we told them to shut it down. Nobody, including me, was prepared for it. They stopped the thing when it stops, it stops on a dime and makes a loud bang and generates huge loads that could cause the film to telescope, but it didn't telescope.
>
> I also remember that Kodak's engineers never had any doubts about the film telescoping. There was a guy named Hans, a German. He had a scaled model of a roll of film, maybe 35 millimeters, about 2 feet in diameter. He couldn't understand why we would need a caging mechanism to prevent the film from telescoping. The more we discussed, it the angrier he got. I told him we had some people who were concerned, primarily Ken McLeish, our vice president, who had some prior experience where they had telescoping problems. But Hans takes this roll, and he got so pissed off that he began banging it on the side of the table, smashing it against the wall—and nothing happened to it! It was like a brick! It took six flights before we pulled the caging mechanism off the supply unit.

Kent Meserve, manager of the Manufacturing department, told a story about another test of the film roll:

> Early in the program we were still in Wilton. I was asked to take a team and equipment to a facility in the Boston area, which had a vacuum tank with windows. This test was requested by the Mechanical Engineering department, headed then by Charlie Karatzas, an amazing man. He was probably one of the world's best mechanical engineers—and one of the worst mechanics.

Anyway the purpose of the test was to take a roll of film (about the same size as a full take-up reel), and determine at what film tension we had to wind film on and off in a vacuum in order to maintain steering and servo control. So we had a rig with a supply, take-up, active steering, tension sensor rollers, etc. We also had a photographer with high-speed movie capability, a servo expert, a couple of technicians, and me. After a couple of days of setting up and checking out of the instrumentation and photo gear, we ran our matrix of spooling tests in both directions as we gradually reduced the pressure in the chamber. We were testing pressure versus controllability.

Eventually we hit the pressure where the emulsion started to outgas. Servos started having problems. Steering of the film was overshooting, etc. Essentially we had an air bar of some 3 feet in diameter! The required tensions in the film to maintain control were getting way too high. And then the inevitable happened—we lost servo and steering control, things went wild, and the film was torn. Before we could dump air into the chamber, the next minute destroyed the supply roll. With no tension, the film was outgassing like mad and the roll started shaking and shimming side to side and whipping layer after layer [of film]. It was a shocking sight. Just like a roll of Jello shaking like mad. You could hear the side-to-side banging through the chamber walls.

We called in the results and headed home. On the way I realized this meant that we needed a pressurized film system. My technical area at that time was path alignment and I did not look forward to having to do it as part of a closed system. As it turned out Charlie and his crew came up with that clever flexure/bracket arrangement, which automatically compensated for misalignments by twisting the film along its neutral axis. In my view, this development ranked right up there with the twister as a key feature of the system. It took care of the problem of the vehicle changing shape due to thermal changes.

Pressurization

The pressurization requirement led to enclosure of the supply unit subsystem in a thin, walled aluminum pressure vessel. Internal pressurization was passive, with the film roll itself supplying the water vapor, therefore achievement of very low leak rates was required to support mission durations. The enclosure incorporated removable ellipsoidal domed covers, which allowed film roll installation/replacement. Figures 6.5, 6.6, and 6.7 show the enclosure, the supporting structure, and the key subassembly arrangement.

Fig. 6.5 Supply unit structure/enclosure.

In order to maintain gas pressure inside the film supply container, a narrow labyrinth path was created for the film to exit the container. This film path consisted of a 4-inch diameter roller having a clearance of only 0.005 inches with respect to a surrounding housing. This configuration, called the film exit vestibule (see Fig. 6.8), created a good gas flow restrictive slot. A silicone rubber lip seal was closed across the film web, thereby assuring no water vapor loss during long periods of inactivity.

A caging mechanism (Fig. 6.9) was installed to prevent any possible axial motion of the film during launch. It was accomplished by actuating a complicated mechanism pushing on both sides of each film roll. Caging was only put into the assembly for the first six launches. After that it was deemed unnecessary and was removed from the design.

Paul Brickmeier remembers a particular incident regarding the caging mechanism.

> Sometimes during the program we had a supply assembly sitting there with the covers off and technicians working on it. Chester Nimitz, our CEO, comes into the area with some corporate board members all dressed up in clean room suits and they lined up around it. Woody Seal, the technician's group leader was tough, could be abrasive, and never feared confronting people. He looks around and says to them, "who the

f–k are you?" and Chester says, "Chester Nimitz." Chester was not offended. And Woody says "oh," and keeps working.

Chester and company came back later into the darkroom through an interlock area after a load of film had been installed and everything was buttoned up and we were going through caging tests. The un-caging process was a violent and loud reaction as it was spring loaded. When the pressure was released it went kaboom.

Woody is outside the room and decides it would be fun to release the caging mechanism. Woody let go the mechanism and everyone leaped backwards about five steps and gasped, as they didn't know what was going on. As they were leaving Woody looks around to find out what was going on and their reactions. He wanted to hear the whole story and I assume to gloat.

Concerning the darkroom we did not wear special goggles. You could vaguely see shadows. You couldn't see anything in particular, only shadows of big objects. You could spot large things and it was all

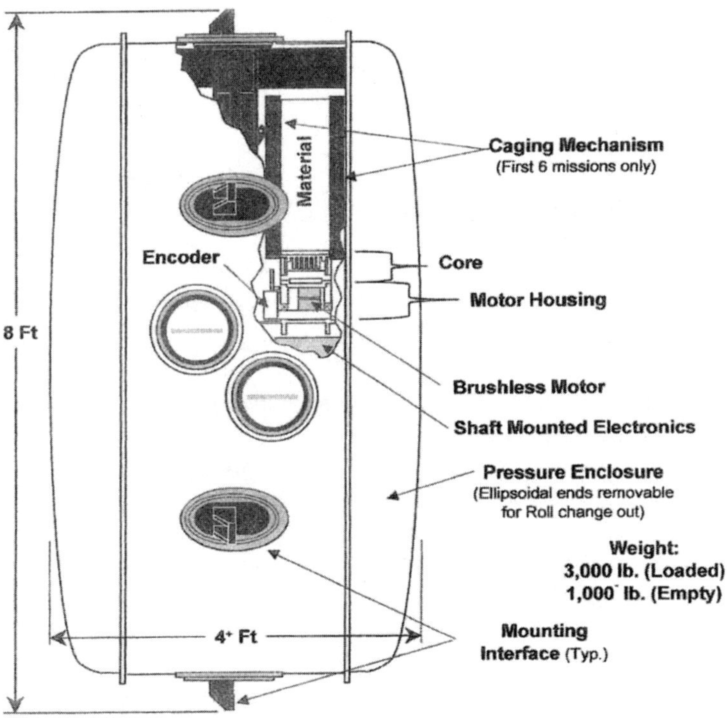

Fig. 6.6 Supply unit partial cutaway view.

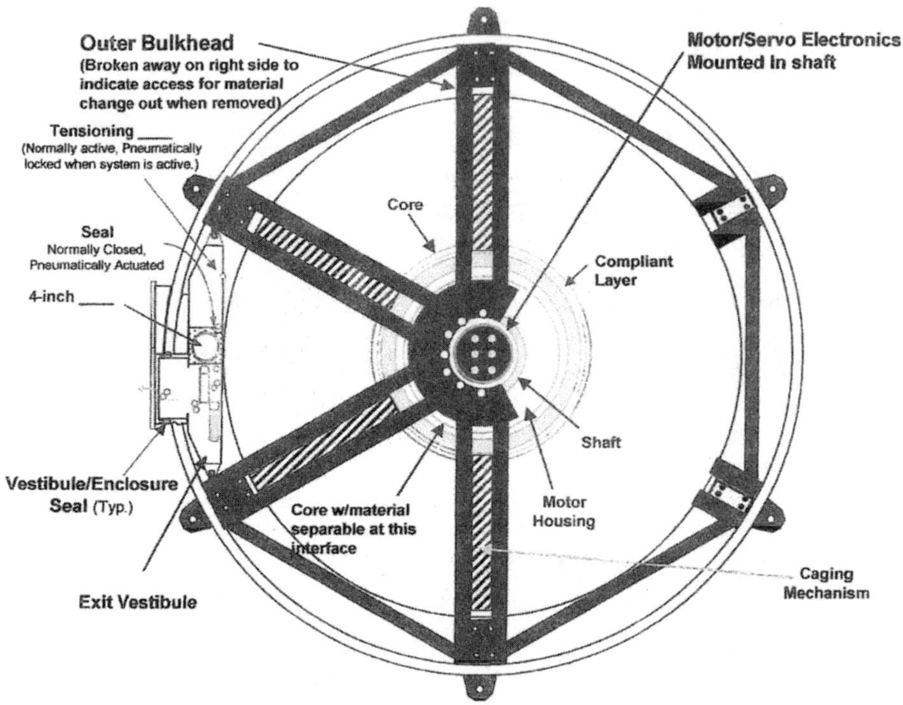

Fig. 6.7 Supply unit with enclosure cover removed; 4-inch roller faces in forward direction.

sensing things and you would move your head because it was all peripheral vision. Central vision couldn't see anything.

Kodak had provided us with some little luminescent pads around the room and they were used strictly as boundary markers so you could sense whether you were in the middle or in the corner of the room. They wouldn't tell us what their wavelength was but there was something about it that you could see but it wouldn't bother the film. The amazing part was that we were dealing with 1000-pound objects hung on a hook from the ceiling and installing it with perhaps a quarter of an inch clearance, and we never damaged the hardware.

It was a manual feel kind of thing. We did a lot of training and we had good equipment. It was amazing how we could function really well without being able to see anything.

It was necessary once to remove the supply covers to fix an internal part. Unfortunately unexposed live film had been installed in the supply assembly that was already also mounted in the midsection of the vehicle

> and located in our high-bay clean room facility. In order to pull the covers off it would be necessary for the whole area to be completely dark. People said they could not make the place dark enough to operate. Jack Shanley, a manufacturing engineer who was a wonderful character and a bull in a china shop, was told to make this place dead dark. He spent 8 to 10 hours working on making the whole area sealed off from other activities and totally dark. Several engineers and one technician then pulled the covers off, did what we had to do and succeeded in total darkness.

Film roll integrity had been assured by Kodak's development of an improved winding machine, extending builder roller technology employed in the paper industry to produce high layer-to-layer internal pressures. A builder roller pressed the film against the previous layer as it was wound onto the core. It could develop radial force on the outer wrap of up to 100 pounds. Our company for its part developed a core of suitable stiffness and strength to support the extremely high pressures at the film rolls' inner radius. The core assembly was comprised of a 30-inch diameter beryllium outer ring, which withstood the winding induced compressive load, and an inner support structure that was fabricated of beryllium and titanium. Compression of the outer core during film winding was isolated from the inner structure by a grooved

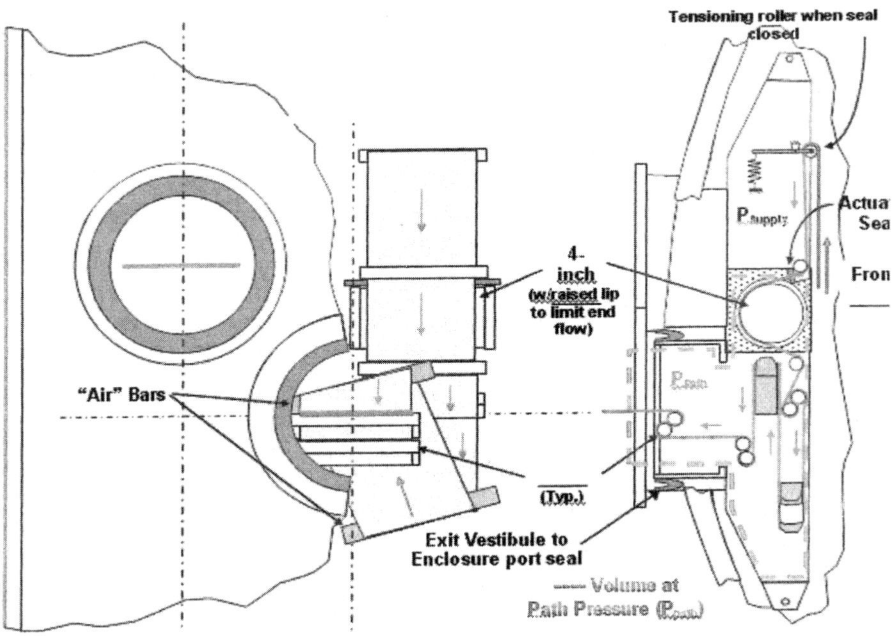

Fig. 6.8 Film exit vestibule.

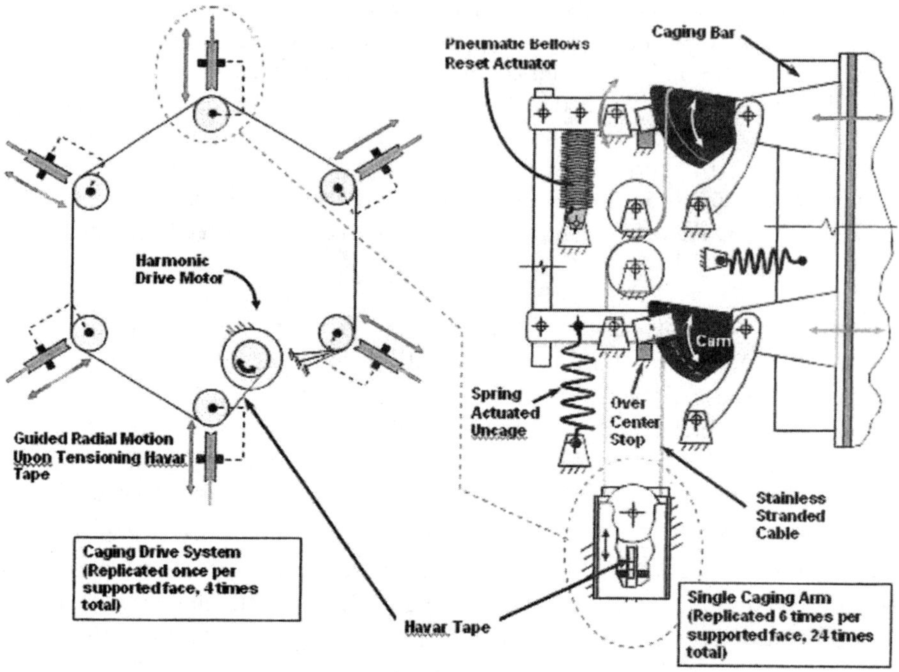

Fig. 6.9 Caging mechanism.

compliant layer of polyurethane. This feature assured that the core inner-diameter deflections were compatible with the requirement of interchangeability and installation/removal from the supply unit subsystem, Kodak's winding machine, and shipping and handling equipment. The empty core weighed less than 50 pounds accommodating one person handling at Kodak. A full core containing 30 miles of film weighed over 1200 pounds.

CHEMICAL DISPOSAL

Processing the film meant chemicals had to be disposed. Bob Jones (Fig. 4.4), director of Engineering, explained,

> We had a lab where we processed film after it was exposed to various optical tests. What happened was when we built the building the developing fluids used during film processing was supposed to go into a holding tank in the building and somehow that got screwed up. The plumbing had it going out into the drainage system and into the pond [just outside the building]. We did that for a while in 1969–1970 and

> then somehow we found out about it, I don't remember how, so we hired some people to dredge the pond because we were scared that it would get into the Danbury sewer system. Even then we were concerned with the environment but it was all kept quiet. We hired some environmental guys and I guess they pumped the pond out and dredged it, and while they were at it, they fixed the dam. The stuff never did go into the Danbury sewer system but I remember that Maguire our program director got yelled at from company headquarters because he spent $35,000 fixing the dam.

The folks at Kodak deserve much credit for the ultimate success of the program. They worked with us in solving some film winding issues and were able to develop thinner and higher resolution films, black and white as well as color. The film path and other key subsystems, namely the looper and pneumatics, are discussed in the next chapter.

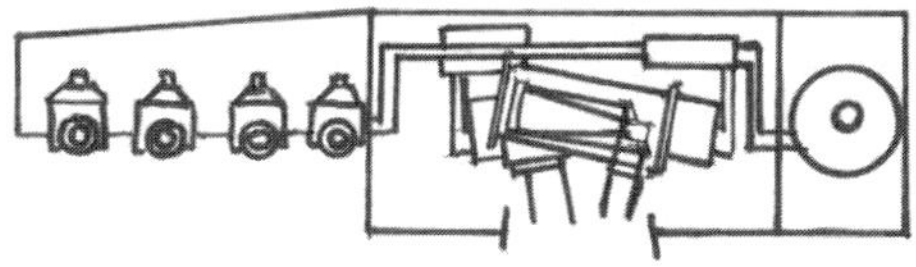

Chapter 7

Looper, Film Path, Pneumatics, and Thermal Design

Ed Butterly, Jr., Dick Carricato, Nick DeFilippis, Joe Prusak, Howard Cummings, Karl Reseck, and Phil Pressel

Besides the film supply unit (Chapter 6) and the camera (Chapter 8), there were a number of key subsystems in the Hexagon. These included the looper and the film path. Like everything else in the Hexagon, they were not simple devices.

Looper

The rotationally oscillating film path in the platen fed film past the focal plane at high velocity and used more film than the supply fed it at a constant lower velocity. It was therefore necessary to provide an intermediate film storage device to accommodate all film speeds and positions in order to get ready for the next photographic frame. This device was called the looper and provided the required reserve film. Nick DeFilippis (Fig. 7.1) was the project engineer on the looper. During the design of the super looper, he was promoted to department manager of Mechanical Engineering.

The looper sounds and in some ways looked like a ride from Disneyland. If it were sized to carry human beings it really would be a wild ride. As complicated as the camera system was, this relatively simple mechanical device was crucial to the system. By providing storage for the film during nonphotographic periods of the rotation of the optical bar, it saved a lot of the power that would have been required to start and stop the system. It prevented momentum disturbances to the vehicle due to starts and stops by commanding film speed changes with a controlled constant acceleration.

The film path for each of the two cameras consisted of two basic systems: the coarse low-speed film path and the fine high-speed film path. The coarse film path, which transported film at the average film usage rate, consisted of the supply assembly that fed film to the looper assembly and then a take-up assembly that received film from the looper. The high-speed film path was in

Copyright © 2013 by the American Institute of Aeronautics and Astronautics, Inc., All rights reserved.

Fig. 7.1 Nick DeFilippis, project engineer on the looper.

the film drive/platen. The film speed in the platen was approximately three times the speed of the film as it left the supply reel.

The best resolution photography occurred at maximum *V/h*, that is, when the vehicle was closest to Earth. This occurred when the film speed going by the focal plane (located in the platen) was the highest. Photography could occur for short scans such as for +/– 15 degrees about nadir (30 degrees total), or as high as +/– 60 degrees (120 degrees total) of the optical bar as it rotated continuously. Each degree of scanning was equivalent to about one inch of film. Therefore a 120-degree scan was 120 inches, or ten feet of film. In addition to the 10 feet of film used for photography, no matter what the scan mode might be, another 2.5 feet of film was required during the acceleration of the film from zero to photographic speed. Similarly, another 2.5 feet of film was required to decelerate back to zero. The total 5 feet of film was run backwards into the looper in preparation for the next frame of photography. These lengths were modified slightly to account for one inch of inter-frame spacing. There was very little wastage of film.

Referring to the 10 feet of film used for 120-degree scans, just imagine the government photo interpreter personnel in Washington examining these incredibly long strips of film for analysis. For stereo, two of those strips had to run together. Until they modified their examining tables and tools, they must have gone crazy.

A schematic of the looper is shown in Fig. 7.2a. For drawing convenience the oscillation carriage is shown moving in the vertical direction. The looper structure is shown in Fig. 7.2b. The looper carriage consisted of eight webs of

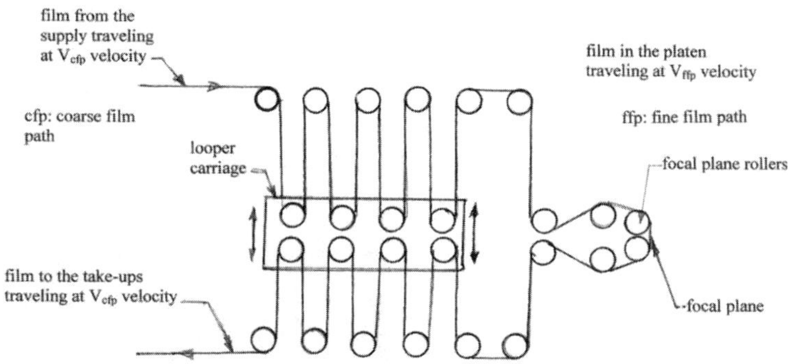

Fig. 7.2a Looper schematic.

film on each side of the moving carriage. The important thing was that the carriage oscillated back and forth receiving film at low speed from the film supply reel on the input side of the carriage while delivering stored film going forward at high speed to the platen during photography. While the carriage travelled in the other direction, it received film coming back from the platen at high speed after photography traveling in the reverse direction back into the looper and then exiting to the film take-up reels.

The film rollers in the looper and throughout the whole film path (except for the air bars discussed in Chapter 9) were made of beryllium and are shown in Figs. 7.3 and 7.4. They were made of beryllium to make them as lightweight as possible and to minimize rotational inertia. The rollers were 0.5 inches in diameter and 7.1 inches long. This was slightly longer than the 6.6-inch wide film.

The all-aluminum structure of the looper consisted of forward and aft uprights mounted to a base. The looper traveled on two hardened-steel round guide rails, which spanned the two uprights. The looper was enclosed in a

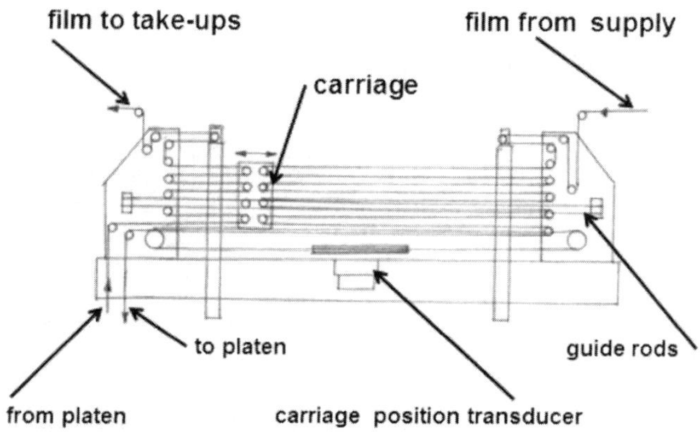

Fig. 7.2b Looper structure.

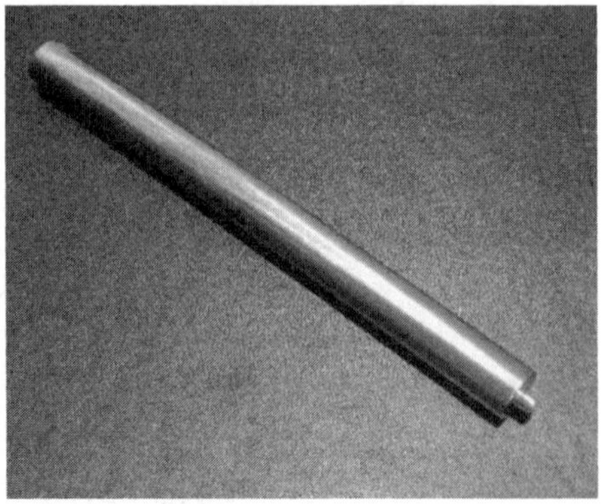

Fig. 7.3 A film roller.

pressure and light-tight enclosure to prevent the film from losing its moisture and being exposed to light.

There were two loopers on the frame, one for each camera. The forward-looking camera's looper was located at the left rear of the frame, and the rear-looking camera's looper was located at the right front of the frame.

> When the program went to a wholly pressurized path, it resulted in some very fancy and expensive containers. The looper was a big box-like design with all of its sides made by numerically controlled milling plates with hundreds of lighting pockets leaving a honeycomb-like structure. Early in the program we were desperate for the delivery of the second container so we authorized the vendor to express ship it by plane from Chicago. We met the commercial flight in New York to pick it up, but no package! They personally delivered it to the airline shipping area so something happened in Chicago. This was before bar code readers so tracking it was done by hand and seemed to take forever.
>
> It took a week to find it. At Customs in Caracas, Venezuela! And they were not cooperative about releasing it. The airline worked it for a few days unsuccessfully and then we had to turn to our customer for help. It took them about another week working with our embassy there to get it

Fig. 7.4 Close-up of the end of a filmxroller.

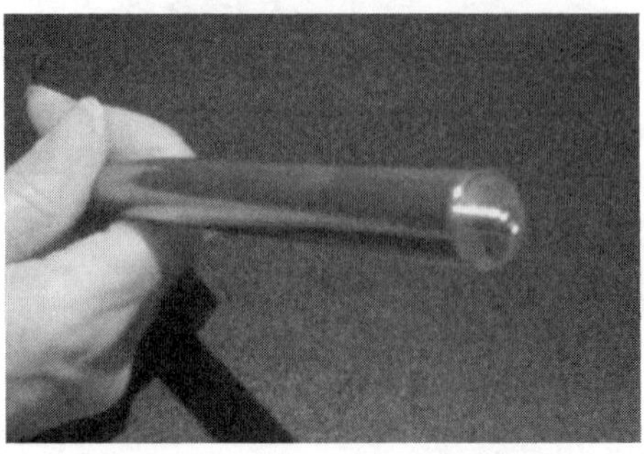

pried loose. From then on, if we were schedule-desperate, we flew someone to the Midwest, rented a vehicle and drove it to our building nonstop.

The carriage was constructed of aluminum and housed eight rollers. Flywheels were fabricated from heavy-met (a very dense material) and sized to inertially balance the carriage as it traveled along the rails. The carriage position sensor assembly consisted of a large disk that drove a potentiometer. The disk was connected to the carriage by a sprocketed steel tape.

During launch the carriage was suspended from an upright against launching forces as high as 30 *g* (10 *g* rms times a factor of safety of 3). It was therefore required that the looper carriage be caged during launch in order to prevent damage to its bearings. The carriage was caged against the forward upright on caging pins and held in place with a two-pronged hook. Power was transmitted to the hook from a solenoid via an over-center toggle arrangement.

Joe Prusak (Fig. 7.5), one of the managers in the Mechanical Engineering department and one of the designers of the film path was sitting in his office one day when Bill Mitchell, one of the vehicle managers from the program office, stopped by. Bill told Joe that an Air Force general was going to visit the facility and that Joe was tasked to help put together a presentation. One of the top items was a description of the design of the super looper that was nearing completion. The super looper was implemented on the system in the early 1980s. Its purpose was to increase the amount of film in storage in the looper in order to permit a rapid response to a command to photograph certain off-axis (away from nadir) targets without having to turn on the supply reels to provide additional film.

As the presentation date approached, super looper project engineer Ed Butterly, who was supposed to discuss it, came down with a terrible case of nausea. The day before the presentation he came to Joe and said that he was too sick to make the presentation and asked if Joe would make it for him. Now what was Joe to do? He spent the next day in a closed conference room

Fig. 7.5 Joe Prusak, Mechanical Engineering manager.

with Ed learning as much of the details of the new design that he could. Ed did an excellent job of explaining the design.

On the fateful day Joe was extremely nervous. He was faced with a general accompanied with his staff including junior officers and consultants from The Aerospace Corporation (known in the technical community as Circle A). Joe felt that all of them would love to nail him with an embarrassing question that would score points for themselves.

How was he going to make this work? He was ready. He said to the general that he would like to describe the design directly from the drawing layout. The general agreed. So he rolled out the layout across the 12-foot conference room table. This cut out all viewgraph participation and limited the dialogue to those at the front table. He took the general eyeball to eyeball through the design. He fielded virtually no questions. Success. Later in the day, his boss Charlie Karatzas, the Director of Engineering (one of our heroes) told him that the general was pleased with the presentation and felt that he knew how it worked. Joe wiped his brow with relief.

Joe also remembered a rather harrowing experience while checking out a system that had already been assembled and was being prepared for launch.

> Several Hexagon systems had already flown when a somewhat scary event occurred to one of our colleagues, Duane Turoczi of the Quality Engineering department. Some of a new lot of stainless steel looper tapes were found to have nicks that could result in fracture during useful life. Duane was sent to Lockheed in Sunnyvale, California, the integrator's facility, to inspect the tapes already installed in the looper. The vehicle (we called it a bird) had been fully integrated and was vertical in a high-bay area. The only access to the looper was on a diving board, positioned several stories high on a scaffold. The diving board protruded through an access door in the midsection. After the looper cover was removed, Duane crawled in belly down on the diving board and began to inspect for damage using a dental mirror. Just then, the area was hit with a so-called minor earthquake that scared the living hell out of Duane. He scampered back off that diving board in record time and climbed down the scaffold to the safety of the floor. It was an experience he will never forget. Luckily he had enough time to inspect the tape and see that there were no significant nicks and he was able to pronounce it safe for flight.

As improvements to the camera system were made subsequent to the initial flights, thinner and better film became available. This also meant more film per flight and therefore more film storage was required. This led to the redesign of the original looper into what became known as the super looper.

This one was larger with more rollers and a carriage that traveled further back and forth. It came into existence in the early 1980s.

FILM PATH

The overall system film path is shown in Fig. 7.6 as a schematic. There were sections of it that were quite short such as in the platen area. Long sections such as between the supply and the looper as well as from looper to take-up were several feet long.

Joe Prusak explains how he started working on the film path:

> As a young engineer in 1967 I accepted a job as a mechanical engineer at the Perkin-Elmer Corporation. I had previously worked there as a design draftsman on the Mercury periscope program while going to school nights. I had also worked at Sikorsky Helicopter and two other technical companies.
>
> I was first given some unclassified work and put into a large bull pen while waiting for my security clearance to come through. When I was finally cleared I was moved to the new plant in 1968 where the program was to be run. I was given a tour of the Hexagon facility and saw models that had been fabricated. There was a wooden mock-up and a metal engineering model. I was also given a written description of the two-camera system and how it was to operate.
>
> I went home that night and wondered whether I had made a good choice by taking this job. As I reviewed what I had learned about this ultra-complicated system, I kept coming up with "they're crazy." Considering the dynamics involved with both cameras that included a constant speed

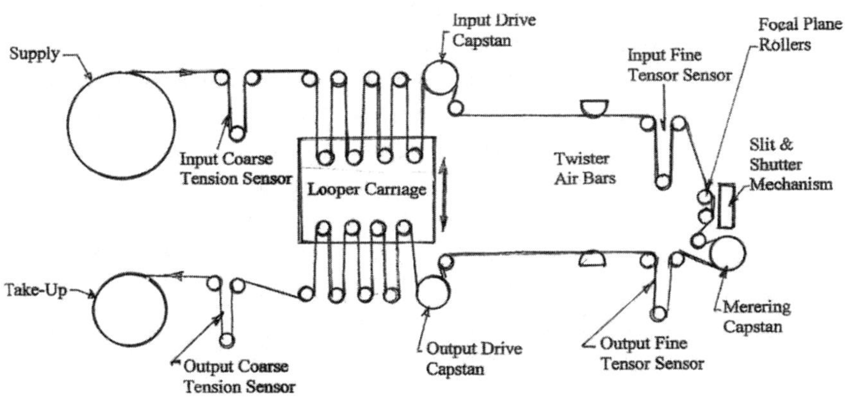

Fig. 7.6 Overall film path schematic.

> supply, a slack box (or looper), and film going forward and coming backward from the focal plane and back to the looper at incredible speeds, I couldn't believe that they could pull it off. It was an amazing mechanical challenge to design, and an even bigger challenge to create an electronic servo control system.
>
> Nevertheless I stayed and much to my consternation I found that they were serious and that I was to be the project engineer responsible for the two-camera assembly film path and interfaces to the supply and take-up.

The flow of the film path was as follows: film exits the film exit vestibule (FEV) in the supply, then enters the looper, then exits the looper and goes to roller 1, then down to the input air bar, then laterally to the focal plane/emulsion-out roller 2, then to the focal plane entrance/exit rollers 3. The film then returns the same way through the other inline rollers and the exit air bars and back to the looper, where it is then sent to the take-ups.

The nominal tension maintained in the film was 2.5 pounds for best film tracking. The long paths were articulated (made flexible) so that dynamic and temperature conditions would not affect the tracking of the film through the system. This included compensating for the residual effects of launch loads and orbital temperature changes affecting the shape or motions of the vehicle and especially the two-camera assembly (TCA). The articulation principle was to keep the entrance and exit film rollers parallel in the direction that the film was traveling while allowing torsional (twisting) motion and changes of length between the rollers.

Reasonably early in the program design phase, problems controlling the steering of the film in sections of the film path occurred. The initial problem was a low-frequency oscillation of the film between the rollers and the air bar when the air bar was disturbed.

Charlie Karatzas (manager of the Mechanical Engineering department) and Bob Jones (director of Engineering), both excellent mechanical engineers, met to discuss the problem. They were stumped and decided to call in Bill Newell and give him a week or so to investigate and hopefully come up with a solution to the problem. Bill Newell was a very senior mechanical engineer (senior in age and in experience with mechanisms and flexures). So they gave Bill the problem and off he went.

Surprisingly he came in the next morning with sketches, analyses, and some recommendations. Problem solved. Actually Bill had done some previous work on articulation and steering and so had not really solved the problem overnight, but that does not in any way diminish his brilliance and creativity.

Bill recommended replacing the air bar steerer with passive, steerable articulators. The big problem was due to a combination of things: the great length of film in the film path, the large number of rollers, the need for recycling the film back and forth, and the relative motion of various parts of the film path. Basically Bill broke up the steering function into two parts. One part was to articulate the film path to permit misalignment of various subassemblies along the film path. The second was to keep the remaining sideward displacement of the film along the film path within a reasonable amount.

Thus the passive articulation provided 6 degrees of freedom in the displacement of any major subassembly relative to the others with the steerer only required to handle 1 degree of freedom involving the side motion of the film. This was implemented and solved the problem. The solution was really a result of the combined efforts and determination of a few, but also on the efforts of many others who obtained information of roller design, flexure theory, alignment techniques, and much testing.

Certain articulators had capacitive devices that sensed wandering of the film to either side of the rollers. The sensors sent a signal to a stepper-motor-driven steering mechanism that drove the film back toward the center of the rollers thus correcting its position. Other articulators that did not provide sensing and steering were passive (meaning not actively controlled by motors).

Because the center-to-center distance from the A side to the B side of the film supply rolls was significantly smaller than the center-to-center distance of the A and B cameras (optical bars), the film had to be transported outward radially in the vehicle to where the photography took place. The crossover assembly consisted of rollers and air bars that served to accommodate this difference. After the film exited the supply it entered the looper, then the crossover that placed the film in-line with the film drive/platen assembly. After photography the film exited the film drive/platen assembly and went back into the crossover, then back to the looper, exited the looper and went into the take-up assembly.

The film exited the looper and passed downward to the crossover assembly that in turn utilized its combination of rollers and air bars causing the film to execute a 90-degree turn into the film drive. Similarly, the film exiting the film drive passed back through the crossover then upward and back to the looper.

On the forward-looking camera the film was delivered directly to the input roller of the looper, as shown in Fig. 7.7. The radial position of the two supplies in the vehicle was smaller than the radial distance between the centerlines of the optical bars; therefore, crossover assemblies were needed to line up the film entering the optical bars.

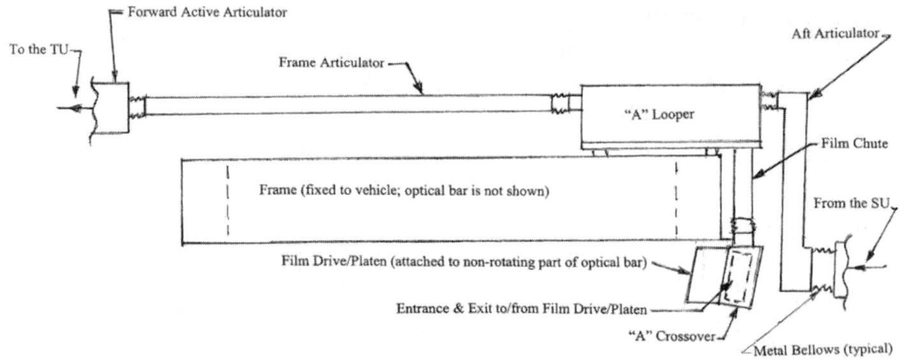

Fig. 7.7 A-side (forward-looking camera) film path.

Upon exiting the looper the film was transported through the frame articulator section to the forward take-up section. A typical passive frame articulator is shown in Fig. 7.8. Both aft active articulators and both of the first active articulators in the forward section also had a stepper motor-driven steering mechanism. The tube and bellows assembly that provided the pressurized film path fit was attached independently to the TCA.

To accommodate relative motion between the enclosures and film chutes, we developed a set of bellows, both elastomeric and metallic. They provided excellent sealing and were soft enough axially to accept differential motions between assemblies so as to not cause additional loading on any parts. The B-side (aft looking camera) path is shown in Fig. 7.9.

The forward film path for each camera used similar components. The first articulator in the take-up section was active and could steer the film in both forward and aft directions. The three articulators between each reentry vehicle were passive. My colleague Glen Grabow was the project engineer

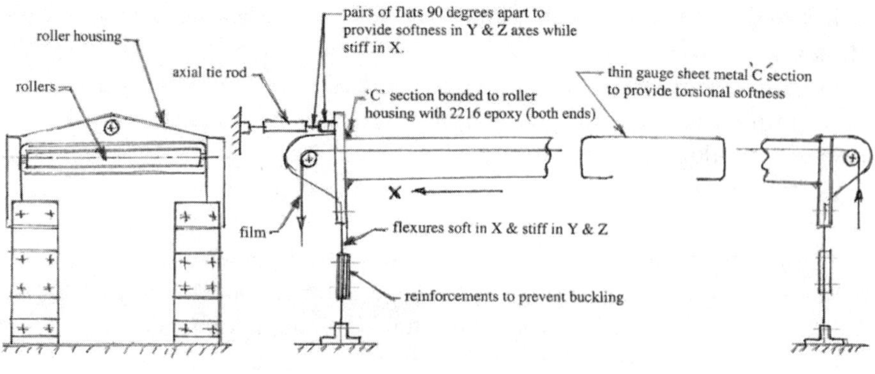

Fig. 7.8 Typical passive frame articulator.

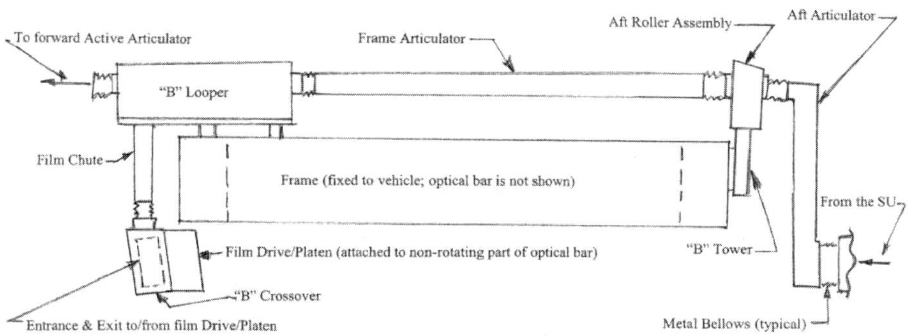

Fig. 7.9 B-side (aft-looking camera) film path.

on the forward section film path within the four take-up reels. Glen's nickname was "Shims" because he tried to solve all technical problems with thin spacers to adjust assemblies.

The overall film path schematic is shown in Fig. 7.10; notice that the two optical bars are missing as the film only goes by the focal plane.

Working with George Seibert, a structural analyst in the Systems Engineering department, we developed a structure that utilized a waffle plate construction for the walls of the enclosures. The waffle construction was to provide stiffness and save weight. The looper was the biggest challenge as it was enclosed in a very large aluminum rectangular box. George thought out all of the loads and stresses very well; then, together we

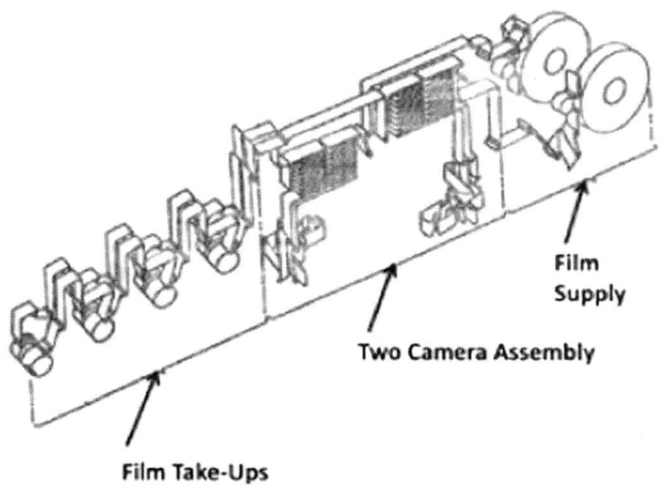

Fig. 7.10 Overall film path isometric.

designed the corner joints to be structurally sound and able to be easily welded.

The whole process went through a very extensive drafting period where we developed George's ideas and the recommendations made by the Budd Company in Philadelphia who fabricated the enclosures. All of the enclosures eventually successfully passed their pressure and leak tests after considerable redesign to improve the quality and appearance of the welds.

When the whole Hexagon system was in vacuum it was also necessary to control the outward pressure in the film path. Two evacuation valves were put in the system and sized to accommodate the worst-case launch pressure reduction rate in order to vent the system. Of course during testing the vacuum chamber pump down rate also had to be well controlled. Two other valves were added to allow for re-pressurization back to atmosphere after testing.

PNEUMATICS

An inert pneumatic gas system was required for the Hexagon system wherever there was film. This included the supply, the take-up assemblies, and the whole film path. The gas was required to keep the film from losing moisture, as the system was in the vacuum of space. The gas was also necessary to keep the film from touching the air bars in the platen and to operate the seal door bellows in the film exit vestibules of the supply. Sealing the film path did several other things: it reduced dendritic discharges and their resultant fogging of the film and provided a back pressure when the supply was unsealed reducing any cooling effects that could lead to snow formation. It also helped keep stray light from the film.

The pressure at the air bars required for optimum system operation was 2.5 pounds per square inch (psi) above the ambient pressure and was set by a pressure regulator. Elsewhere in the film path the gas pressure was maintained at 0.7 psi. The gas used was 99 % pure dry nitrogen.

The pneumatic system consisted of two identical and independent sides, one for camera A and one for camera B. Each consisted of a pressure tank, several shut-off valves, and a pressure regulator. The two units were interconnected via a crossover shut-off valve on the high-pressure side of the regulator (near the tanks) and a low-pressure crossover valve on the low-pressure side. This interconnection provided system redundancy. A schematic of the pneumatics system is shown in Fig. 7.11.

The tanks were spherical and made of Maraging 200 steel (Fig. 7.12). This material was chosen for its extremely high strength. The tanks were approximately 25 inches in diameter and were pressurized to 3265 psi with

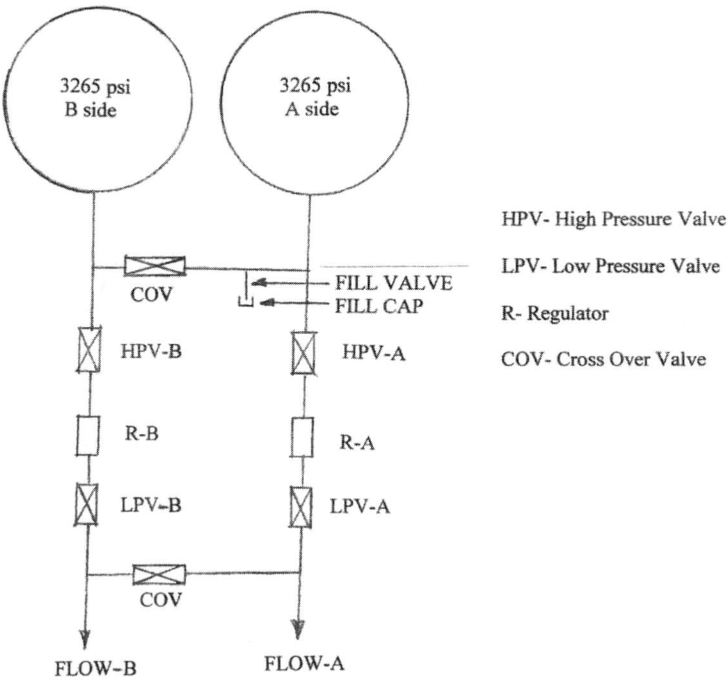

Fig. 7.11 Pneumatics schematic.

the nitrogen. This was sufficient to provide the Hexagon system with enough gas for its maximum projected mission length.

It was critical that the tanks were properly made and tested because at over 3000 psi they acted as bombs capable of causing enormous damage. Fortunately during the life of the program there were no failures of any kind in the pneumatics system.

Thermal Design

Changes in temperature and thermal gradients can distort and/or misalign optical components. Therefore it is critical for a high-performance camera or telescope to have its temperature well controlled.

The temperature of the Hexagon system was maintained at 70+/− 23°F during operation and 70+/− 30°F during storage and nonoperating conditions. These numbers were chosen to prevent severe impact on the optical performance. The relative humidity was always controlled to between 45 and 55 percent to maintain the proper moisture content in the film.

In order to provide such a tight temperature range there had to be a sophisticated thermal control system. Thermal control was done passively

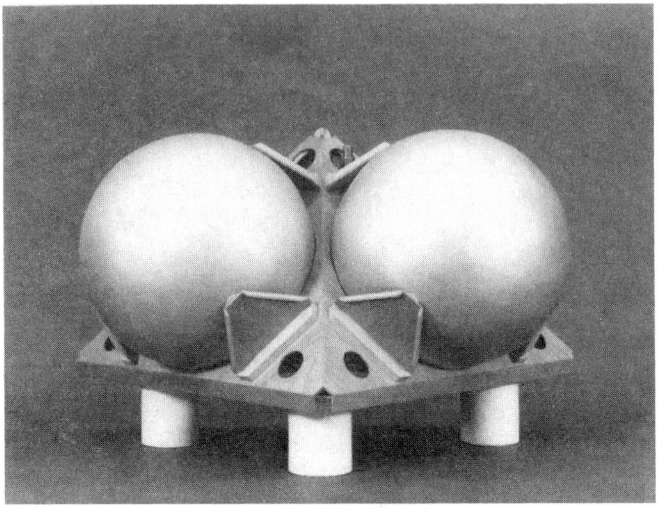

Fig. 7.12 Pneumatic tanks.

rather than actively. This meant that there were no heating or cooling elements in the system. For example, all of the heat generated by electronics was conducted and radiated away. The heat from the sun impinging on the spacecraft and albedo (the sunlight reflected from the earth onto the spacecraft) as well as the loss of heat due to exposure to the ultra cold of deep space was all controlled by the thermal design. It was basically a "heat balance" problem.

For example, the optical bars were essentially enclosed in a thermal cocoon. They looked out the open entrance viewport around which was a fiberglass baffle. The baffle was used to help control the temperature in the area and to prevent stray light from entering the optical system. The open entrance was protected during launch by a fairing that was discarded just prior to orbital insertion. During launch and times of nonoperation the optical bars were stowed inward facing the area away from the aperture. This area was also covered with insulation.

A thermal control system requires analyses ranging from simple hand calculations to sophisticated computer modeling. There are so many variables in creating a good thermal control system that thermal testing of the eventual hardware was an important part of its design and usually led to necessary changes. The thermal design also had a direct effect on the overall system's weight, size, and complexity.

The internal surfaces of the surrounding structures were black. This reduced stray light and thermally minimized enclosure temperature gradients. Generally if components were black, it was to radiate energy away (like from electronic boxes). The polished surfaces on structures like the optical bar tubes and the frame resulted in low emissivity and would decouple them from the environment and minimize response to transient fluctuations in temperature. Heat was also rejected using structural panel radiators. Some components were coated with a black coating having a high emissivity, meaning it could easily absorb heat by radiation.

Portions of the camera system and the vehicle were enveloped with flexible blankets of multilayer insulation (MLI). MLI is made up of layers of thin (on the order of 0.001-inch thick) aluminized Mylar or Kapton plastic sheet materials. Each layer was separated from the next with a net-type of separation material such as Dacron to prevent direct thermal contact between the plastic sheets. The layers were sewn together at the edges and in some interior areas to form blankets of MLI. Perforation holes were required at many places in these blankets in order to allow trapped air to escape into the vacuum and prevent ballooning of the blankets during ascent into space. The Lockheed staff that dealt with thermal analysis and control of the vehicle thermal environment deserves much credit for coordinating and maintaining all the thermal interfaces with the various contractors.

All camera mathematical thermal models incorporated only radiation and conduction heat transfer mechanisms. Convection was regarded to be negligible and was omitted. Standard modeling techniques were used in which hardware was represented by a network of nodes (elements), with each node representing a finite portion of hardware. Surface properties of nodes (coatings, size, and shape) control the amount of energy transferred by thermal radiation to and from a node. The mechanical properties of nodes (thermal conductivity, heat capacity, size, shape, and mass) control the amount of energy transferred by conduction. Temperature of a node results from the net energy exchange for that node, in both temperature level and temperature transients. A node has a spatially uniform temperature (over its entire surface area and its entire volume).

Three levels of modeling granularity were used for the Hexagon camera, which can be roughly characterized as coarse model, detailed model, and very detailed model. Coarse models had relatively large nodes, with large surface areas for radiation and long paths for conduction. More detailed models had smaller nodes with smaller areas and shorter conduction paths. To model specific hardware, the number of nodes increases as the level of modeling detail increases.

During the preliminary design phase, a total program interface-definition working meeting was held in a large building in Los Angeles. Participants represented our customer and all contractors. Participants were allocated to functional working groups for the purpose of establishing a preliminary definition of contractual interface parameters. The thermal working group included engineers from Perkin-Elmer (main camera) and Lockheed (spacecraft system and vehicle support compartment).

Representatives from Perkin-Elmer and Lockheed met to define the contents of an interface control document to establish environmental interface temperatures. Lockheed was responsible for supplying hardware that would be the thermal environment to the camera. Based on prior analyses Lockheed would provide 1) an insulated enclosure (cocoon) dedicated to the TCA and supply assembly, and 2) the baffles outside the viewports to help shield camera optics from the external environment. The interface definition included examples of orbital parameters, camera operating profiles, camera power requirements, solar flux, albedo (reflected energy from the earth) ranges, Earth infrared temperature, etc.

It was a joint Lockheed/Perkin-Elmer requirement to maintain the camera's two optical bars, the supply assembly, and the four take-up assemblies within the required temperature range during operation on-orbit. In addition, Perkin-Elmer had the implied requirement of maintaining a spatially uniform temperature throughout each optical bar in order to achieve contractual optical performance. The optical bar, which I worked on, is described in the next chapter.

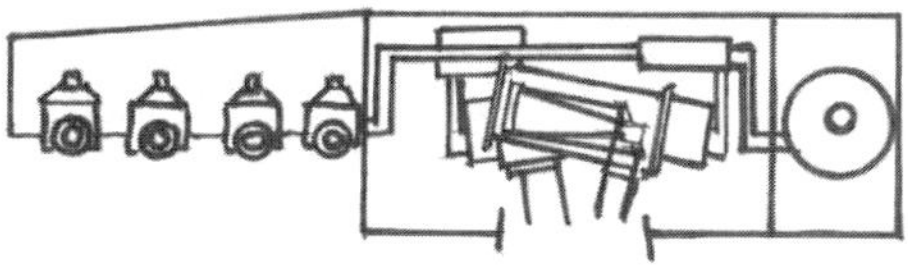

Chapter 8

Optical Bar

Phil Pressel

Configuration

The optical bar was the light-gathering assembly of the Hexagon camera. The term *optical bar*, in the field of optics, means a camera or the primary optical assembly.

Two optical bars comprised the two-camera assembly (TCA) (See Fig. 8.1). The forward-looking camera was on the left side of the vehicle as it moved forward. This camera was designated camera A. Camera B was on the right side and looked aft. This arrangement was used to obtain the stereo effect. Both optical bars were attached to the frame.

The frame was the structure holding all of the camera's major mechanical components except for the film supply and the take-up. The two optical bars held by the frame are shown in Fig. 8.2. The top, side, and end views show the positions, direction of rotation, and stereo angle of the two optical bars. Figure 8.3 depicts the configuration of either of the two optical bars, which were identical.

The diameter of the optical bar was 30 inches, its length approximately 6 feet, and it weighed about 600 pounds. The optical bar was an f/3 camera. The f number of an optical system is the ratio of the lens focal length to the diameter of the entrance pupil. It is a quantitative measure of lens speed. The focal length was 60 inches. Light entered through a round glass corrector plate whose purpose was to perform minor optical corrections to the incoming light. Prior to entering the 20-inch diameter corrector plate, the light passed through a fiberglass viewport baffle whose purpose was to prevent stray light from entering the camera (such as reflections off of the spacecraft) as well as to control heat transfer in the area.

The optical bars rotated continuously but in opposite directions during photography. Figure 8.4 shows both optical bars from the front with the pneumatic spheres in the background. If an observer were looking from the front or rear of the vehicle at the two optical bars, each would be seen to

Copyright © 2013 by the American Institute of Aeronautics and Astronautics, Inc., All rights reserved.

Fig. 8.1 The TCA, comprised of two optical bars.

rotate in opposite directions similar to an old-fashioned clothes wringer (where clothes dropped between the rotating bars would be squeezed as they dropped through as per Fig. 8.2). The bars rotated in opposite directions to minimize momentum disturbances to the vehicle. Figure 8.5 shows a side view of the optical bars tilted with respect to each other and the pneumatic tanks adjacent.

A counterweight was mounted on the support tube opposite the glass corrector plate in order to keep the optical bar rotationally in balance. The

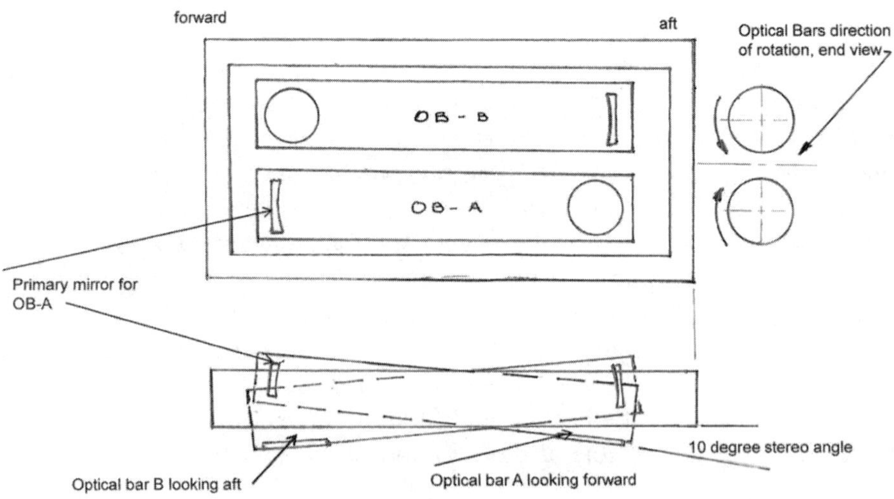

Fig. 8.2 The optical bars in the frame.

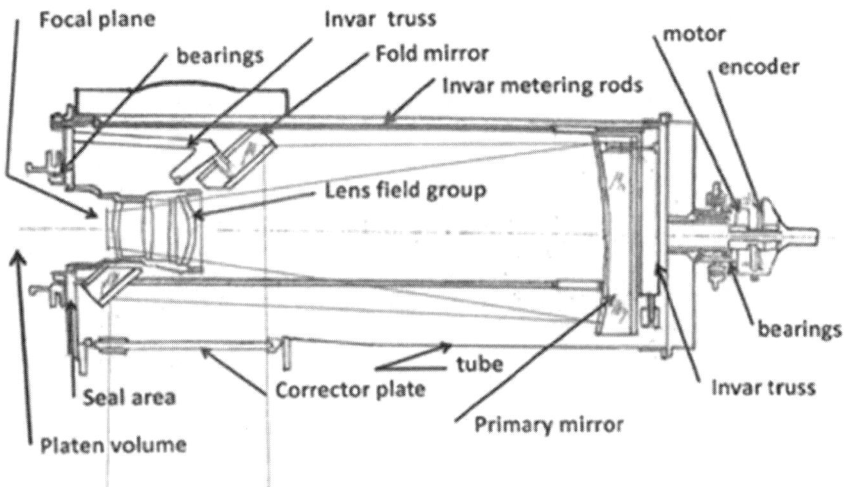

Fig. 8.3 The optical bar.

counterweight consisted primarily of an electronic box called the digital data acquisition systems. It transmitted signals through slip rings to the electronics in the vehicle.

After the light passed through the corrector plate shown in Fig. 8.6, it was reflected off a rectangular fold flat mirror that was tilted at 45 degrees to the corrector plate. It directed light to the round primary mirror. The mirror focused the light through an assembly called the field group, located within

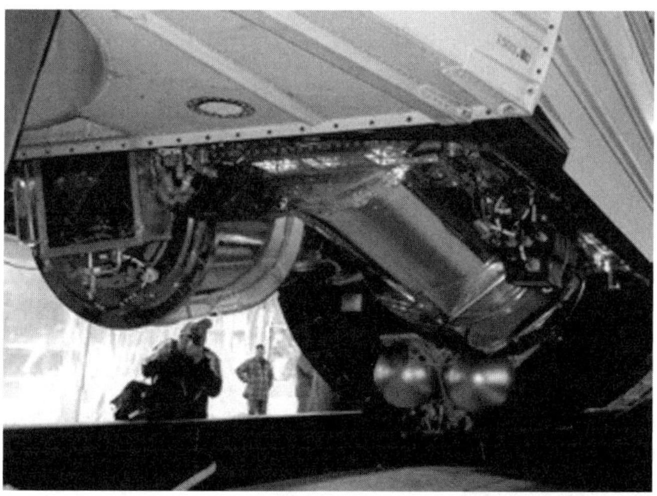

Fig. 8.4 Both optical bars (front) with the pneumatic spheres (background).

Fig. 8.5 A side view of the optical bars tilted with respect to each other and the pneumatic tanks adjacent.

the central hole of the fold flat and was comprised of four lenses. Eventually in 1974 an interchangeable optical filter was added to the field group assembly and was located between the second and third lenses. It was used when new types of film were introduced into the system.

So many of the engineers on the program—whether they were systems engineers, mechanical, or electrical engineers—were incredibly talented. In my opinion, the three best engineers that I worked with were Dick Carricato, Joe Vollaro, and Tony Bassaro (Fig. 8.7). The reason is that they all paid particular attention to details, knew the fundamentals of their specialties, and were thorough and successful in everything they did.

Fig. 8.6 The corrector plate and platen area.

Fig. 8.7 The three best engineers I know: Dick Carricato, Joe Vollaro, and Tony Bassaro (photographed at an October 2008 reunion).

Dick Carricato, an electrical engineer, was intimately familiar with how the whole complicated system worked. He helped develop the equations of motion for the film, optical bars, loopers, and tension sensors. He worked on the servo feedback system that accurately synchronized the film speed and position to the speed and position of the image in order to minimize blurry photographs. He also was called upon to solve on-orbit problems.

Joe Vollaro, a mechanical engineer, was usually called upon to solve or design some of the most complicated mechanisms. He dabbled in everything, designing the optical bar interchangeable filter, mirror flexures to minimize "bumps" on mirror surfaces, platen capacitive transducer glass plates, and platen stops—plus many other components. Joe solved many problems, often with elegant solutions.

Similarly Tony Bassaro, also a mechanical engineer, designed some of the more complicated mechanisms in the system. He had the ability and the tools (at home) to actually machine some of the parts of his designs himself and try them out before committing to a final design.

FOCAL PLANE

The interchangeable filter consisted of two glass filter plates mounted perpendicular to each other in an L-shaped frame and driven by a motorized mechanism. The focal plane was located 1 inch behind the last lens, called the back focal length (BFL). The focal plane's actual position was in the platen cavity (see Fig. 8.3). Originally the BFL was only 0.5 inch but there were so many mechanisms and electronics between the focal plane rollers and the rear of the last lens (such as a shutter and electronic boards) that I asked the optical designer at the time for more room. The designer was an old-timer named Lloyd McCarthy, gauntly thin with a long gray beard, who barely responded to me with a grunt. Two days later he called me in and quietly said "you can have 1 inch." I replied "thanks" and left.

With respect to the focal plane, Len Farkas, a systems engineer and eventual program director talked about data received at the focal plane.

> Film was—and in many ways still is—able to capture far more information than any digital imagery system [used] today. The Hexagon camera ran film through the focal plane at 200 inches per second. It was almost perfect with minimal amount of smear. That put 30 billion elements, or pixels, on film per second. That didn't even include the levels of intensity, which are measured with a microdensitometer. When that is taken into account, it adds up to about one trillion bits of information a second. Today there is no means of storing that amount of information.

Similarly, Bob Jones, our director of Engineering (Fig. 4.4), talked about information. "The information-gathering capability was staggering. I went down to NPIC [the National Photo Interpretation Center in Washington, D.C.] once to give a pitch to the photo-interpreters about our progress. 'You know, you better get ready, because you're gonna see more information in an hour than you've ever seen in your life,' I said to them."

Since Len and Bob made their observations, scientists have found a new way to store data. The European Bioinformatics Research Institute (BEI) in England has found a way to encode genetic material (DNA) with data using a breakthrough process that stores huge amounts of information. Currently, however, it is much too expensive a process for industry to use.

The fused silica primary mirror was supported kinematically (three-point mounting) with flexures attached at three equidistant points on the mirror in the plane of its center of gravity and at their other end to invar mounting trusses. This mounting design was used to minimize bending loads from being applied to the mirrors. *Invar* is a steel material with a very low coefficient of expansion (0.7 millionths of an inch/inch per °F) and close to that of fused silica (0.3 millionths of an inches/inch per million per °F).

Three *metering rods* held the axial position of the primary mirror, relative to the focal plane. These rods consisted of a long portion of invar and a short length of aluminum. The aluminum length was set so the metering rods had an "effective" coefficient of expansion that kept the primary mirror's axial position constant with respect to the focal plane over the expected temperature range.

The axial position of the primary mirror needed to be held constant to approximately 1 micron in order to meet the precise performance requirements over the temperature range. The operating temperature range of the optical bar was controlled to 70°F+/−30°F.

A micron is one millionth of a meter, or 0.000040 inches. If the rods supporting the primary mirror were made of aluminum, the 1-micron positional accuracy could not be maintained because aluminum's coefficient of expansion is very high—12.6 millionths of an inch/inch per °F.

This meant that, for example, for every 10 degrees of temperature increase, a piece of aluminum 1-inch long would expand by 0.000126 inches. This appears to be a very small number; however, for the approximately 50-inch long metering rods and for a 30°F change, the equivalent axial motion of the focal plane would be 0.019 inches. This was much higher than the requirement. Thus the low expansion invar was required for most of each metering rod's length.

In order to save weight, the 0.75-inch diameter invar rods were designed to be hollow. Due to their length and need to be straight, the hollow rods were gun-drilled. *Gun-drilling* is the process of drilling a long straight hole in anything, such as a rifle, that is so long that ordinary machining tools such as lathes cannot achieve.

The key person who designed the metering rod assembly and method of temperature compensation was Roy Stoll, a thermal analyst who was a major contributor to the thermal design of the overall system. (Unfortunately Roy died in 2004 and could not contribute to this book.) The thermal design is discussed in Chapter 7. It was also necessary for the optical bar's main support tube and end bulkheads to be highly buffed and polished to a shiny surface in order to control its temperature.

Talking about small numbers, Bob Jones (Fig. 4.4) remembers that many from Perkin-Elmer thought that even though the SETS personnel from TRW were supposed to provide technical oversight, they lacked the knowledge to do the job. He remembered the time that the SETS experts pored over test data indicating that the film would rise off the rollers by about a micron. Ray Anderson, an Air Force colonel, was chairing a meeting in which Bob said, "Ray, I want to show you something. And I pulled a hair from my head and held the hair up. I said, 'This hair is a hundred microns in diameter. Will you tell them to shut up?' Anderson stopped the meeting just like that. The SETS guys were in another world."

The primary mirror's lateral position was held by three tangential links tying its support truss to the tube of the optical bar. The links had spherical bearings at each end in order to prevent bending loads from being applied to the primary mirror support structure.

On the focal plane end of the optical bar, a stiff bulkhead was the mounting surface for the metering rods, the fold mirror, the lens housing, and the rotating part of the platen. On the other end of the optical bar was situated its drive motor, encoder, slip rings, and supporting vibration isolator. Inasmuch as slip rings are bad for use in space because they can generate debris particles and electrical noise, the slip rings on the optical bar were used for nonmission-critical thermal diagnostics only. Any conceivable failure mode in the slip rings would not degrade performance or system life.

The platen, discussed in Chapter 9, was the assembly that fed the film at controlled speed past the focal plane. The film position needed to be precisely at the focal plane within millionths of an inch. The velocity of the film had to coincide exactly with the rotating speed of the optical bar. This speed ranged from 10 to 30 revolutions per minute (rpm) depending on the ratio of the velocity of the vehicle divided by its height above Earth. The photographs with the best resolution were at the maximum ratio, namely closest to Earth, when the optical bar speed was 30 rpm and the corresponding film speed in the platen was 200 inches per second.

SUPPORT

Both ends of the optical bar were supported on pairs of angular contact bearings. The bearings on the focal plane (platen end) end of the optical bar were 14 inches in diameter. The bearings on the drive end were 4 inches in diameter. The stationary housings at both ends were attached to the frame. The bearings were extremely accurate. Their preload served several purposes:

- To minimize axial looseness in the bearings
- To prevent axial vibration loads during launch (from “lifting off” bearing balls from their races and chattering), thus preventing *brinelling* (generation of small indentations made by the balls when highly loaded against the surface on which they are running; for a highly precise instrument such as this camera, such small marks can cause significant torque disturbances and thus affect performance)
- To control the predicted bearing running torque
- To control the stiffness of the bearings

The retainers in the bearings that held the balls in place were made of a porous plastic material called phenolic. The bearings were lubricated with a spaceflight-rated oil made by the Braycote Company. The retainers were impregnated with the same oil and served as an oil reservoir during the mission. Because the bearings were exposed to the vacuum of space, a narrow labyrinth gap was designed in the retaining rings to hold the bearings in place. The labyrinth served as a barrier keeping a significant quantity of the oil from escaping into space through capillary migration or evaporation (called *outgassing*). The labyrinth also kept contamination from entering the bearings.

There was one instance when there was a bearing problem in the assembly and test area. It was a particularly humorous episode that occurred early in the program's development work on the engineering model testing. The assembly had spent quite a lot of time in vacuum chambers and at one point the optical bar bound up and would not turn freely. It had been built up with off-the-shelf, non-space-qualified ball bearings.

Charlie Karatzas, our department manager, had come into the clean room to review what was going on. After being briefed on the binding-up problem, he correctly diagnosed it. The moisture in the phenolic ball retainer ring had evaporated in the vacuum and caused the retainers to shrink and bind up the bearings. Luckily we had access to the bearings. Charlie instructed one of the assembly technicians to fill a hypodermic syringe with water and inject the water into the bearing at several places in order for the retainer to absorb the water. The technician was not about to inject water into the bearing (a highly unusual thing to do) but he was also not about to tell Charlie to get lost, so he hemmed and hawed for a while. When Charlie turned his back, the tech sneaked away. Charlie got a syringe of water and injected it himself. Half an hour later the bearing worked fine. No one else could have come up with such a simple and unique solution. Bravo Charlie.

In order to prevent any vibration forces in the launch direction from creating any structural deformations near the focal plane, it was decided to support all of the axial launch loads on the optical bar at its end furthest away from the focal plane—namely the motor/encoder end. At the focal plane end, there was a set of three kinematic tangential links tied to the frame. At the motor end, the optical bar was supported by a stiff vibration isolator tied to the frame. The isolator reacted to both axial and lateral loads of the whole optical bar. Both the links and the isolator were designed to control the vibration frequency response of the optical bar to prevent it from being "excited" by the launch vehicle vibrations. The optical bar lateral loads were reacted through by the isolator on one end and the three tangential links on the other.

Composition

Many parts of the optical bar were made of aluminum. Aluminum is lightweight and easily machined and fabricated. The bulkhead located at the focal plane end was made of beryllium. This bulkhead supported all of the components that determined the position of the focal plane, specifically the two mirrors and the lenses. Beryllium was chosen for this bulkhead because it is an extremely stiff material, lighter than aluminum but just as strong. Beryllium has the highest stiffness-to-weight ratio of any metal. This stiff structure could therefore ensure that the focal plane would maintain its position at all times.

One might ask, "Why not fabricate the whole optical bar out of beryllium if it has such nice mechanical properties?" The main reason is that it is extremely expensive. If, for example, aluminum were to cost $1 per pound, beryllium would cost $1000 per pound. It also requires special processing and machining precautions. Due to its toxic nature, it is important to avoid inhaling beryllium powder generated during the machining process.

Stress analyses were performed on all components of the optical bar when undergoing maximum loads. These acceleration and vibration loads occurred during launch. As stress analysis was a forte of mine I did my own analyses. This was an advantage because to meet our design review schedules, I had to come up with designs in a timely manner and did not have time to wait for the back-up analyses done later by more sophisticated computer stress and dynamic analyses by specialists in the Systems department. My results were always pretty close. I usually designed hardware to have positive margins of safety based upon the precision elastic limit (also called *microyield*) of the materials. This was done to ensure that there would be no residual deformations (or misalignments) due to material yielding as a result of launch loads.

Mirrors

The Herraeus Corporation, located in Germany, fabricated the mirrors. Graham Wallace, my colleague and a fine opto-mechanical engineer, was sent to Herraeus to discuss the technical issues concerning the mirrors (weight, material, and configuration) and their ability to meet our cost and schedule requirements. They agreed to supply us with the primary and fold mirrors for the duration of the program.

The mirrors were made of fused silica and consisted of front and back face sheets 0.5-inch thick and a similar material called rotosil as the 3-inch core. Their sandwich construction configuration is shown in Fig. 8.8. The purpose of the core was to create a stiff structure that would support the polished front

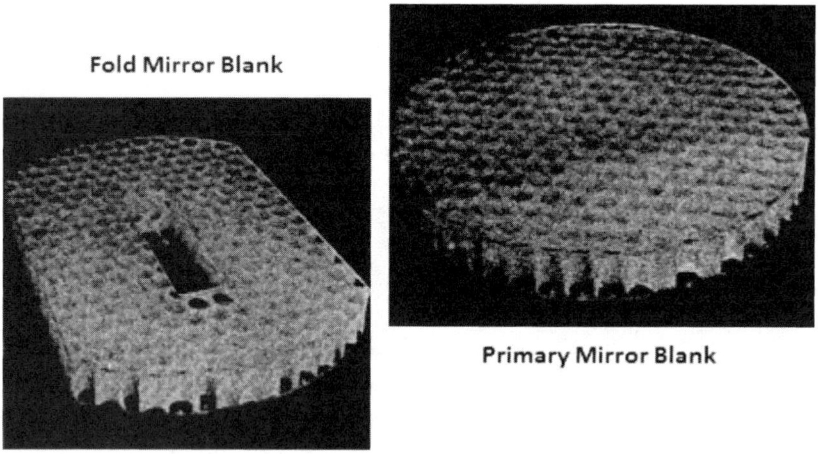

Fig. 8.8 Mirror blanks.

faceplate. The core had many holes drilled through it to create a lightweight structure. It was then fused under high temperature to the front and back faceplates. The mirror blanks were inspected via an ultrasonic technique to ensure complete fusion.

When we received the mirror blanks we polished them in our Wilton, Connecticut optical shop. The fold flat mirror was polished to a root means squared (rms) surface finish of one fortieth of a wavelength of light. The unit *rms* equals the average height of the optical surface discontinuities compared to the maximum and minimum heights of discontinuities. The primary mirror was polished to one fiftieth of a wavelength and the lenses were polished to one twentieth of a wavelength of light. Just to give an idea of what that means, a mirror with a fiftieth of a wavelength of light surface quality is flat over its entire surface to approximately half a millionth of an inch. That is really flat.

Kinematic mounting of the circular primary mirror meant supporting it at three points because three points define a plane (see Fig. 8.9). Four points supporting a mirror creates deformations in the mirror because one of the four points will not lie in that plane, thus creating bending in the mirror. Lyman Spitzer, Jr. was a world-famous astronomer working at Princeton. He was the person who originally, in the mid-1940s, conceived of the idea for a stargazing telescope orbiting the Earth in outer space. Today he is considered the father of the Hubble Space Telescope. In 1966, I obtained a copy of a paper written by Lyman Spitzer on the design and methods of supporting any large circular mirror. I decided to call him on the telephone to ask for his ideas on the design for our 26-inch diameter mirror (much smaller than the eventual 8-foot diameter Hubble mirror).

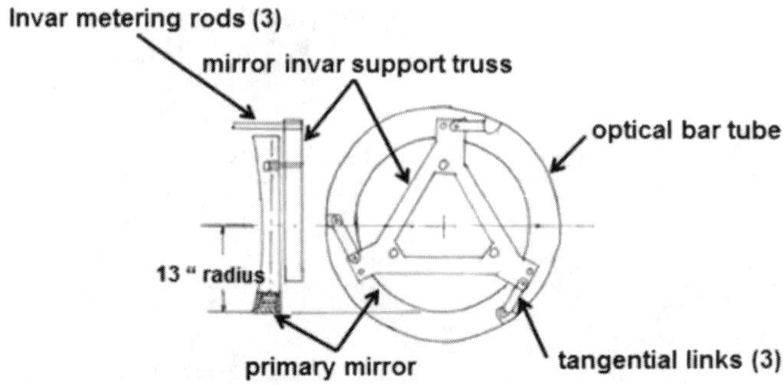

Fig. 8.9 Primary mirror support.

The question was where to support the mirror. Would it be at three equally spaced points on its edge or somewhere in the middle along a radius line? If it was supported at the edges, then the center would have the maximum sag due to Earth's gravity. The goal was to support it at some intermediate points on the radius to minimize sag. Lyman gave me some good advice. He recommended that the three points supporting the mirror be located at a radial position located at 60 percent of the radius of the mirror. Back in the 1960s computer finite element analyses were not yet developed, so the locations were determined by experimentation. We fabricated a circular 0.25-inch-thick aluminum plate with the same diameter as the primary mirror.

The vertical sag of the plate was measured at many points with a dial indicator. The plate was then loaded with a known quantity of sand held inside edge barriers and the deflection of the same points was then measured. The sand represented the weight of the mirror. This procedure was repeated for support points located at various radii. The optimum location of the support points was found to be at close to 70 percent of the outer radius. That is the location used rather than Spitzer's 60-percent recommendation. The same sand measurements procedure was repeated for a plate simulating the rectangular fold mirror. Compared with today's finite element computer analyses, this was pretty crude but it did the job.

Unfortunately the fold mirror needed to be held at four points instead of three, due to its rectangular shape. In order to create an artificial third point, two of the support points on the mirror were tied together using a crossbar. It turned out that the crossbar created serious astigmatism problems in the fold mirror when the mirror position was on edge with respect to gravity, thereby creating a bending load. After months of studying this problem, a special *tiger team* (engineering jargon referring to a special team established to solve a problem) recommended a redesign of the crossbar.

PRESSURIZATION

The discussion of the film path in Chapter 7 explains that it was necessary to prevent moisture evaporating from the film. Experiments showed that the evaporation was minimized when the whole film path was sealed and pressurized with dry nitrogen. As seen in Fig. 8.3, there is a path in the pressurized platen/optical bar cavity where the nitrogen could escape through the optical bar 14-inch bearings into the vacuum of space. It was necessary to design a seal to keep the nitrogen contained in this area. The problem was that the escape path of the nitrogen gas was between the stationary and rotating portions of the optical bar. Therefore a rotating seal was required—not an easy design task.

The final seal design is shown in Fig. 8.10. After much testing with various configurations and materials, the final design consisted of two sheets of

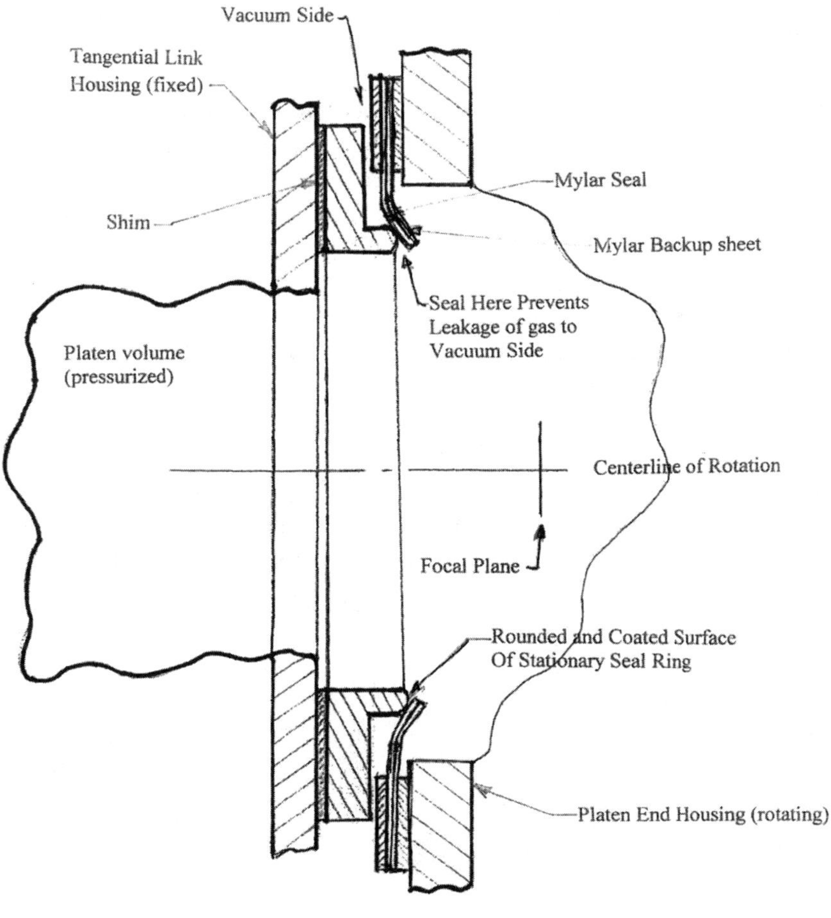

Fig. 8.10 Optical bar rotating seal.

Mylar (a polyethylene with an ultra high molecular weight). The edge of the front sheet rubbed against a finely polished hard, anodized aluminum ring. The face of the ring was coated with a special lubricant made of a molybdenum–disulfide compound called "Lifelube." This coating was to minimize the friction torque. The back sheet was used to control the stiffness of the seal.

Testing consisted of measuring torque and leak rates on a simulated rotating seal pressurized with nitrogen and small amounts of helium. Helium was used because it can be identified with a leak detector.

When the torque spikes were detected prior to launch, it was decided prudent to clean the rubbing face of the Mylar. I performed this myself several times at Lockheed after the whole TCA was installed in the vehicle. While the vehicle was vertical I had to lie down on a sort of diving board near the top of a 50-foot scaffold that reached into the vehicle. I was able to see the seal directly from one direction as well as by looking through a fiber-optic bronchoscope threaded through a lubrication access hole.

I always managed to remove the residue, but I did have one very bad experience when performing this task. I was lying down on the diving board when a clamp that I was using fell from my grasp. Unfortunately I had forgotten to tether it to the board. It fell approximately 50 feet down to the bottom of the aft section of the vehicle. This was the section where the supply assembly resided. Luckily the supply had not yet been installed and the clamp was visible. Had the supply been there, the clamp would most likely have fallen behind it and been unreachable. This would have necessitated the removal of the supply assembly in order to get to the clamp. Fortunately technician Jim Walls, who was working with me, was clever. He soon located a magnet and a roll of lacing cord. We "fished" for the clamp with the magnet and after a short time snared it and gently raised it to the safety of our hands. It was an embarrassing moment; never have I told anyone about it until now.

Occasionally similar torque disturbances were transmitted back to Earth on several orbiting vehicles. I recall being called in the middle of the night by the person on duty in our war room. I came to work and analyzed the data charts. In each case the torque signatures were similar to previous ground test data, and I reassured the vehicle manager and the division general manager that the torque disturbances were acceptable and that the camera would continue to operate successfully. It did each time. Good news—in orbit it could not be repaired.

INTERACTIONS WITH OTHER DEPARTMENTS

My department manager was Charlie Karatzas, a real gentleman. But I interacted directly quite a lot with Arnie Wallace, who was on staff with upper management in the Program Office. Arnie was a cigar-smoking hard-working person. Unfortunately, he suffered a heart attack during the early years of the program. As he recovered he still proved valuable to the Hexagon program, entrusted with the charter of ensuring the program's success. Specifically, this meant he had to confirm that the program was technically sound and met its requirements so that Perkin-Elmer would earn its maximum fee.

As such, one of his tasks was to challenge many of us on our designs. He was very knowledgeable and leaned hard on us to prove that our hardware would perform properly. He looked at all parts of the system and deeply questioned the respective responsible engineers on why this or that would or wouldn't work. I viewed him as our program's devil's advocate. His job was to represent the customer and make sure that any technical anomaly was addressed as to whether it was a one-time occurrence or a generic problem and to ensure the high reliability of success while maintaining schedule.

I believe that he did this job very well. He was insistent on getting answers and would not let go until he got them. Many of my colleagues did not care for him because he could be pushy and critical, but his forcefulness resulted in many of us finding flaws in our designs and fixing them. As I worked with him over the years, he grew to be my hero even though many on the staff thought he was a troublemaker. In the end he did save us from some key mistakes, thus sparing the program much grief. I believe that without Arnie's efforts we might have made mistakes that would have made the program less successful.

Perhaps some did not appreciate his special tie clip. It was a miniature revolver that actually worked by detonating a cap. One time I was looking over a design in the drafting room when he came by and looked over my shoulder. There was something about the drawing or some comments made by another draftsman that he did not like. The next thing I heard was this extremely loud explosion that shocked all of us. I turned and looked the tie clip revolver in his hand while he said, "You freakin' guys, get on the ball." He did not use the revolver often, but he frequently called whoever he was in a meeting with "you freakin' guys." Freakin' guys was one of his favorite set of words to use if he was not satisfied with whatever was going on. He never used curse words.

Many problems required presentations at the West Coast customer shop. This task often fell to me with Arnie in the audience. The trick was to present the problem, analyses, testing, and conclusions knowledgeably and confidently. But the biggest requirement was to know when to shut up and let Arnie take over from the floor. Much to our regret Arnie died a few years ago. Thanks, Arnie, for all of your contributions.

Joe Zelle worked in the Design and Drafting department, was in charge of standards and specifications, and worked on the drafting board. His claim to fame was that he played the clarinet and in his youth had been a bandleader. His band used to play nightclubs in New York and Connecticut. They also used to play at high school proms and played at a preparatory school prom in 1960 that my wife had attended. He also produced a Christmas show every year for the children of Perkin-Elmer employees. The actors and musicians in the show were employees. It was always a good show and starred Frank "Stumpy" Bulsom. He played the part of Junior, a funny wiseguy kid who used to get in trouble. Frank was a humble and wonderful guy.

I took a few trips with some of our Purchasing guys. These were trips to vendors who machined many of our mechanical parts. On one trip to Philadelphia to visit the Budd Company, which was fabricating the tubular shell of the optical bar, two purchasing people accompanied me. One was Marsdin Shaw, a quiet and reserved gentleman. The other was Steve Sprindis, a loud, boisterous, and hard-drinking veteran of World War II. We were having dinner in our nice hotel's dining room, and Steve was on his third martini while telling war stories. He had been an army sergeant stationed in North Africa. One night he was on guard duty when his unit was attacked by some German troops. He punctuated his story by waving his arms as he recounted the shooting and loudly simulated the gunfire by yelling "rat tat tat tat." Of course most of the people at surrounding tables turned and stared at him. Marsdin and I would have loved to disappear from the table, but in retrospect it was quite funny.

One of the Manufacturing supervisors on the floor of the assembly areas was a real character named Woody Seal. Woody used to make fun of Hans Haberland, one of the Manufacturing department managers. Hans had a heavy German accent. He had been in the German Wehrmacht during World War II and had been wounded, resulting in him getting a wooden leg. When Woody was annoyed at something Hans did, he used to threaten to nail his wooden leg to the floor. This was not well received by Hans, who did not have a sense of humor, but the rest of us laughed privately.

Other departments were involved in the designs we created in Mechanical Engineering: Quality Assurance, Reliability, Electrical Engineering, and Systems Engineering all needed to sign-off on final drawings. Some of the people involved had memorable names such as Dick Fink and Gene Aha in Electrical Engineering as well as Bill Schnitzel in Quality Assurance. My nickname was "Phil Pretzel" sometimes. Ed Brelling used to kid around when we had sign-off meetings that the key signatures were Schnitzel, Pretzel, Fink, and Aha.

The departments also celebrated together. It was traditional after each design review for the various major subassemblies for the director of

Engineering or the general manager to give a party later that evening for the participants in the reviews. This included all of the engineering disciplines. Most of these were in our cafeteria and were wonderful events. There were always plenty of shrimp cocktails, canapés, roast beef and even liquor, wine, beer, and soft drinks. A few times the parties were held at restaurants. We all appreciated the parties that management sponsored to acknowledge all of our hard work.

The next chapter describes one of the most complicated subsystems—the platen.

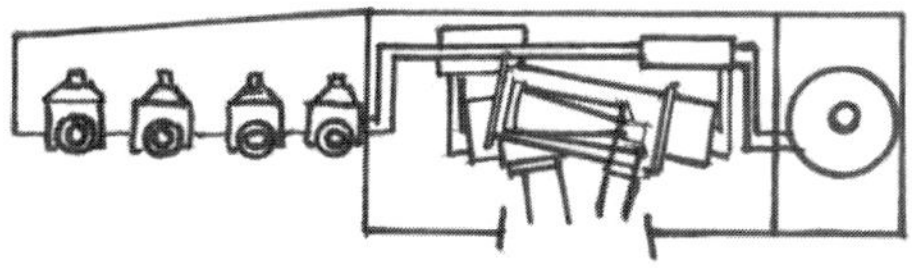

Chapter 9

Platen and Film Drive

Dick Carricato, Bob Hufnagel, Werner Pollack, Joe Prusak,
Ken Stout, and Phil Pressel

Each major mechanical subassembly of the Hexagon system and its associated electronics were important. This chapter and Chapter 11, dealing with the servo-control system, describe the most complicated portions of the system and the most difficult to explain. The platen was an incredibly complex and densely packaged mechanical assembly. It drove and placed the film at the exact position and at the exact speed required to produce excellent photography.

During a photographic scan, the film's rotational speed in the platen had to coincide exactly with the rotational velocity of the image at the focal plane. Also, the film's linear velocity was dependent on *v/h* and on the scan angle.

The film supply and take-up reels were large and had a lot of mass. If one were to start and stop their rotation for each scanning exposure, a large amount of precious electrical energy would be consumed. In addition, the rotational accelerations and decelerations could impart considerable disturbing torques to the spacecraft. Thus, it was desirable to keep the film reels in continuous motion during a run of photographic scans. However, the film needs of the scanning camera were quite intermittent. Actual photography was made only during that part of the camera's rotational cycle while it was looking downward and scanning 120 degrees or less. Therefore, in order not to waste film during the nonphotographic portion of the optical bar rotation, the moving film was temporarily stored. It was looped between a zig–zag series of rollers whose separations were dynamically and continuously changed to match the exact amount of film needed to be stored at each part of the camera's rotational cycle. Essentially, the film needed to be rewound up to the end of the last picture frame and stored until the next photo cycle.

The storage device had two sides: the supply side that stored film from the supply until it was called for a photographic run and the take-up side that stored the exposed film until it was discharged to the take-up reels. The storage device was called the looper and is discussed in Chapter 7. The total

Copyright © 2013 by the American Institute of Aeronautics and Astronautics, Inc., All rights reserved.

length of the rapidly moving film path was quite long in the looper and in much of the rest of the system film path. In such long paths of fast-moving film, small lateral disturbances could quickly magnify and become disastrous. The film could run off from the ends of the rollers and air bars. Thus, lateral film motions had to be controlled. Part of the solution to maintain passive film lateral alignment came in the form of mechanical linkages holding each air bar and roller parallel. These articulators are discussed in the Chapter 7 (see Fig. 7.8).

A means had to be provided to simultaneously accommodate the rotation of the film support mechanism and permit linear motion of the film past the focal plane. This was done with devices called the twister and the air bars. The twister consisted of two air bars and a set of idler rollers. One air bar was for the input film to the focal plane and the other for the output film from the focal plane.

AIR BARS

The film path in the system requires several 90-degree turns of the film. One might think that a simple way to do this would be to orient a roller at an angle so that the film would roll off in a different direction as shown in Fig. 9.1.

Upon closer examination however, the surface of the roller always moves perpendicular to the roller axis but the film does not. That is to say, the film would scrape the surface of the roller as it moved by. Scraping that would result in scratches on the film was unacceptable. Film tension would also be unequal side to side and could cause a crash as the film jammed against fixed hardware.

Fig. 9.1 Skewed roller.

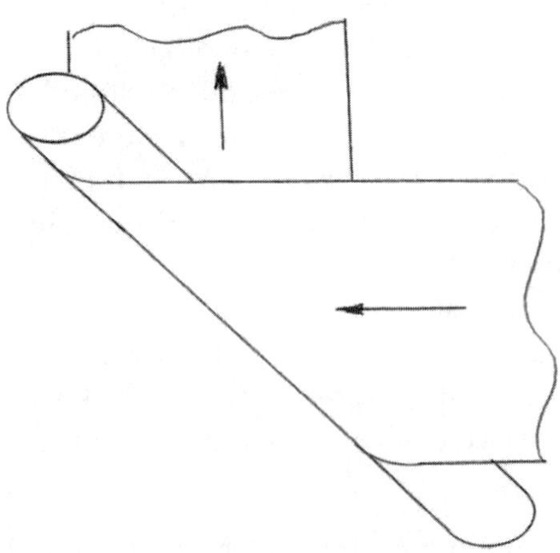

Fig. 9.2 Bob Hufnagel, scientist.

A solution to this problem would be to prevent physical contact between the film and the roller. It was hoped that a flow of air between the roller and the film would prevent such contact, and thus tiny holes were drilled in the roller and pressurized air was made to flow under the film. This did not work.

Bob Hufnagel (Fig. 9.2), one of our talented scientists, recognized that the problem was in the shape of the roller bar. The peak gas pressure is at the apex of the film bend. The outward laminar flow of the gas, constrained by the narrow gap between the film and the air bar, results in a pressure drop along its outward flow path. When the film leaves its proximity with the air bar, the gas pressure must become zero in the vacuum of space.

The local film bending radius is inversely proportional to the local gas pressure. Because the local gas pressure depends on its path position, there must be a careful interactive balance between the local gas pressure and the local air bar curvature.

Bob achieved this balance by calculating and developing the optimal shape of the roller that successfully prevented film from contacting the roller as it slid by. He also specified the size and required positions of the holes in the air bars and achieved the additional goal of minimizing the consumption of gas. The air bars were hollow aluminum bars shaped like the letter D (see Fig. 9.3). The curved part of the D was the surface where Bob placed the

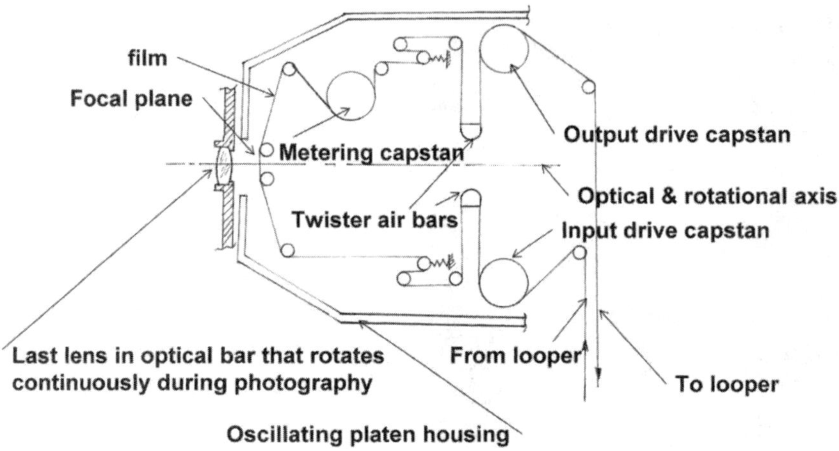

Fig. 9.3 Schematic of the twister; notice the D-shaped bars.

holes to permit nitrogen gas to flow through, thus creating the needed layer of gas between the air bars and film.

There were numerous holes. The diameters of the holes were on the order of thousandths of an inch. The gas used was dry nitrogen. During operation there was a swishing sound coming from the platen. This was the sound made by air escaping through the air bar holes as the film went by the air bars while they oscillated with the twister.

The whole film path, from supply to take-up, was pressurized with dry nitrogen at 0.7 pounds per square inch. As such, the gas was also force-fed into the air bars. The nitrogen gas prevented the film from touching the air bars, except during the time that the film was *not* being transported through the camera, meaning when the film was stationary during nonphotographic periods of the orbit. The nitrogen gas in the supply assembly kept the water content in the film. Prevention of the escape of water in the supply was the "sealing" of the exit of the film through the narrow labyrinth path of the film exit vestibule (FEV) in the supply assembly. In the rest of the film path, the film outgassed water to its partial pressure that was determined by temperature. The dryer film in the rest of the path was already exposed and would be moved to the take-up assembly without going through the rigors of photography.

Twister

The development of an ingenious device called the *twister* was the invention of Don Cowles (see Appendix B). Don was a mechanical engineer whose contribution of the twister combined with the air bars was the

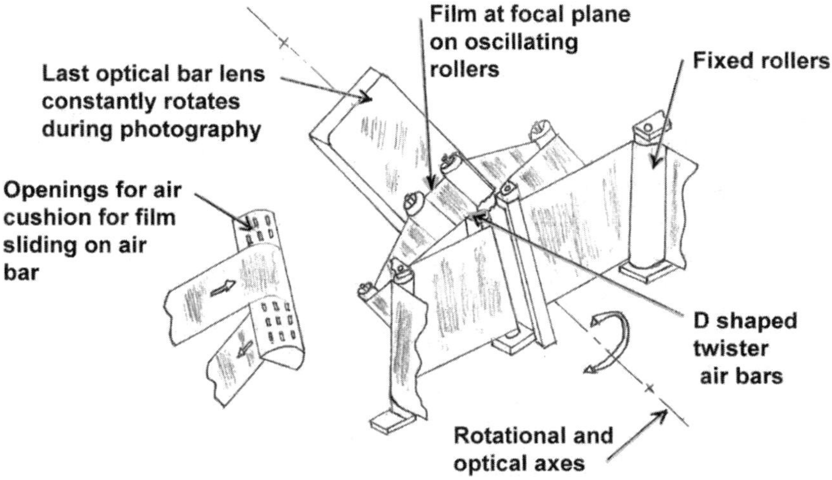

Fig. 9.4 Isometric line drawing of the twister.

innovation that made this program so successful. The twister was the mechanical device that allowed a short length of film from within a contiguous supply to be rotated during photography while still moving linearly very rapidly. The configuration of the twister and air bars can be seen in Figs. 9.3 and 9.4.

FILM DRIVE AND PLATEN ASSEMBLIES

The film drive and the platen assemblies shared space at the focal plane end of the optical bar. The film drive contained an input drive capstan (motor) that pulled film from the supply path and fed it to the platen in synch with the demands of the servo system. An output drive capstan pulled film from the platen and fed it to the take-up film path after exposure.

The film drive also consisted of several idler rollers and a structural housing. The housing was attached to the stationary part of the platen, which in turn was attached to the optical bar's nonrotating support housing. The drive capstans were designed as modules so they could be removed and replaced if problems were encountered during ground testing.

The platen consisted of fixed, rotating, and oscillating parts. The rotating part of the platen, which was attached to the rotating portion of the optical bar, contained the metering capstan that was used to control the speed and position of the film extremely accurately as it sped by the focal plane. The drive capstans in the film drive were slaved to the metering capstan in the platen. The capstan housings as well as the platen and film drive structures were made of beryllium to take advantage of its stiffness and lightweight properties. All of the rollers (except those in the twister) in the system—and there were hundreds—were made of beryllium for the same reasons. It was advantageous to minimize the weight and therefore the inertia of the many rollers. The rollers were hollow and their external diameter was slightly larger than half an inch. Their end caps contained small bearings that were pressed on at each end.

The platen oscillated in synch with the rotation of the optical bar during photography for 120 degrees or less depending on the commanded scan angle about nadir (i.e. +/− 60 degrees). It oscillated on a pair of 9-inch diameter bearings located between it and the optical bar. The film had to speed by the focal plane in a linear motion (forward during photography) and in rotation at exactly the same speed as the focal plane that rotated with the optical bar. The speeds of the film and of the image in the optical bar had to match within 0.02 inches per second (an incredible achievement) for the best resolution at a film speed of 204 inches/second when the satellite was closest to earth. During nonphotography (the remaining 240 degrees or more) the platen rotated in the reverse direction and the film direction also reversed and reached 204 inches per second.

The air bars themselves did not rotate about their own axis or oscillate with the platen. The air bars only oscillated about the rotational axis defined by the optical bar and platen bearings. The concentricity and location of the rotational axes had to coincide very closely (within 0.0002 inches) with the optical line of sight of the optical bar.

The oscillating portion of the platen drove the rollers back and forth in rotation about the axis defined by the platen bearings. The two idler rollers in the platen located at the focal plane were positioned close to each other to prevent a large unsupported span of film. As a matter of fact, the film at the focal plane needed to be held flat to within 4 to 10 microns across the 6.6-inch film web. A micron is 40 millionths of an inch, so 4 microns was 0.00016 inches. The useable portion of the film that was used for photography was 6 inches out of the 6.6-inch-wide film. The remaining edge space was used to deposit data (such as timing marks and position over Earth).

The focal plane idler rollers positioned the film at the theoretical focal plane as defined by the optical bar. Also packaged in the small volume of the platen was a blade-type shutter that let light impinge on the film during a photographic run. The timing of the slit opening of the shutter was a function of *V/h* and the type of scan being performed. Typically the shutter took milliseconds to open or close.

IMAGE MOTION COMPENSATION

Image motion compensation (IMC) is very complicated and crucial for good photography. It is probably the hardest concept to understand because it has many contributors, some of which are quite subtle.

The resolution of a frame of photography from a scan is affected by relative motion between the focal plane and the aerial image. This is called *smear* and is equivalent to someone moving a camera while taking a picture of a stationary object, or a stationary camera taking a picture of a moving object, or both.

Hexagon's platen rotated in sync with the optical bar during 120 degrees of rotation (or less depending on the scan commands) when photography occurred. The platen was then recycled during the remaining 240 degrees of optical bar rotation.

The major component of image motion across the focal plane (along the length of the film) was caused by the scanning of the optical bar. The film had to be moved very precisely to prevent smearing the image on the film. Highly accurate optical position encoders were developed for the optical bar and the metering capstan to ensure this accuracy.

It would have been great had this been the only source of image motion, but it was not. The next image motion, in terms of size, was that caused by the vehicle traveling over the ground. The explanation for the other sources of image motion is described in Appendix C.

FOCUS CONTROL

The film's position on the focal plane rollers had to be adjustable in the axial and tilt (pitch) directions in order to make the focal plane coincident with the aerial image and to adjust for any residual tilt of the rollers. During final ground thermal vacuum operations, through-focus tests were performed in order to initially set and calibrate the axial position of the focal plane rollers. The focus position, from calibration photography taken at the beginning of a mission, was verified after the first of the four reentry vehicles was returned to Earth. It turned out that, for all missions, the preset focus position was the optimum position and the focus control did not have to be used again during the operational phase of the missions. Missions had an engineering run that included a through-focus operation, similar to the thermal vacuum tests on the ground, to determine where best focus was. Of course the results of this test could only be evaluated after the film in the first reentry vehicle (RV-1) was returned.

Fig. 9.5 Werner Pollack (on right [Phil Pressel on left]), designer of the focus and tilt mechanism.

The focus adjustment mechanism controlled axial motion in increments of 2 microns (0.00008 inches) and in tilt of 0.04 arc seconds (0.0000114 degrees), incredibly small numbers. Werner Pollack designed the focus and tilt mechanism. Werner was originally a farm boy who developed into an enthusiastic and fine engineer (Fig. 9.5).

SHUTTER

There was a shutter in the platen consisting of two beryllium blades that opened and closed by sliding laterally next to each other very rapidly. They were activated electrically. A stepper motor set the slit width of the shutter for the required scan angle. This setting controlled the exposure time. The slit was opened and closed normally approximately one milliseconds at nadir, and in less time at other scan angles. In case of a power failure, a once-per-revolution platen cam drive cocked and released the blades; however, this could only set the exposure for a 120-degree scan.

For each photographic frame, the time and the ground location had to be identified on the film. This information was inscribed optically onto a small section near an edge of the film during a fraction of a second stop of the film after exposure. The information was transmitted by means of fiber-optic light pipes. There were two of them in each platen, one being a backup. Because the platen oscillated back and forth, the light pipes (Fig. 9.6) entering the platen were made in a loop long enough so they could follow the same path as they lay in grooves located in the platen support housing. They were designed to survive a minimum of 100,000 cycles as this was the expected mission life.

Sometimes a minor part could create problems and delays. One time, for example, just prior to a launch there was a partial failure in one of the light pipes. There was no time to disassemble the platen from the camera assembly. It was decided to replace the light pipe from the back by removing the film drive. All the screws were removed except for the last one, which

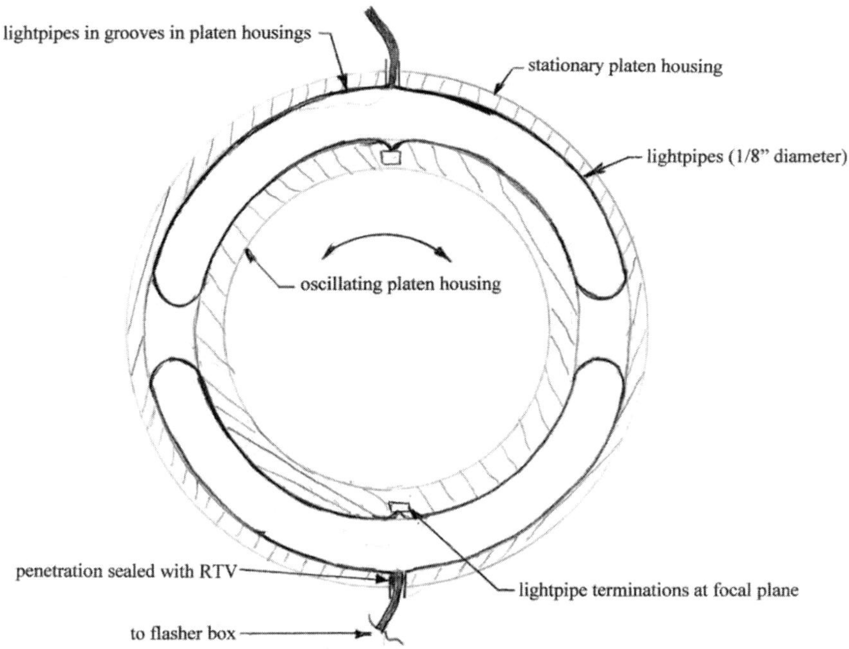

Fig. 9.6 Light pipes in the platen.

was located where the wrench would interfere with the support frame. Technician Duane Turoczi spent hours retrofitting the wrench until finally it worked. It was a Sunday and after seven long hours, with Arnie Wallace looking over his shoulder making it even longer, the job was done.

In the next two chapters the other most complicated subsystem, the servo-control system, is described.

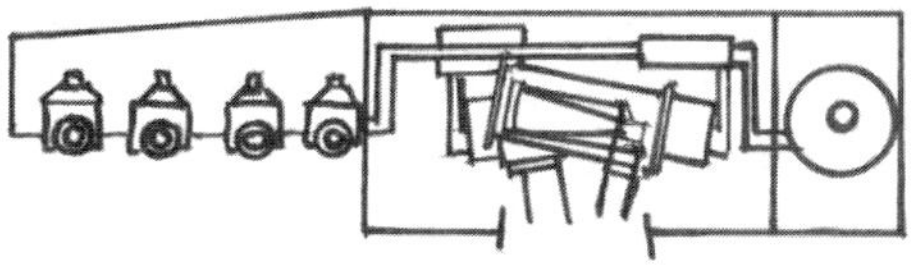

Chapter 10

JOINING THE CORPORATION

DICK CARRICATO*

It was 8:20 a.m. on June 15, 1965, a beautiful summer morning, and I (Fig. 10.1) was on my way to start a new job. My drive to the Electro-Optical Division (EOD) of Perkin-Elmer took me along a route where indications were obvious that this suburban residential neighborhood was changing to light industrial use. But there were still many stately homes surrounded by towering oaks and maples that spoke of a past elegance and a reluctance to give up to industry.

The EOD building did not detract from the beauty of its surroundings. Its blue-green brick blended in with the trees that surrounded the building. The front of the structure gave the appearance of being all glass behind its many pillars. The result was that it looked much more like college campus library than a manufacturing plant. It was within these academic surroundings that I reported for my first day of work. I didn't know it at the time, but it was exactly six years to the day of the first launch of Hexagon.

This was the first time I had ever reported for work not knowing what I was supposed to do. The only instruction I had received was to report to the Personnel department. There I was presented with the usual bank of forms to be filled out, including a U.S. Department of Defense (DOD) personal security questionnaire (PSQ). At the time I wondered why I had to fill out the PSQ as I already had a DOD secret clearance. I imagine that the information in the PSQ was used to enable the background check required for the special program clearance.

After lunch I was introduced to several engineers who would be funneling work to me and to others like me, who were waiting for clearance to the

Copyright © 2013 by the American Institute of Aeronautics and Astronautics, Inc., All rights reserved.

* In this chapter electrical engineer Dick Carricato tells of how he started at Perkin-Elmer and explains the servo-control system. Dick's expertise of servomechanisms, or control systems, made him the go-to guy to investigate and determine the causes of significant anomalies prior to launch and while operating in orbit. As Len Farkas, former director of the program, says, Dick "was one of the most important and least appreciated and most honest people on the program. That guy would do anything to help you out....if it wasn't for Dick Carricato we would not have been so successful....He was close to a genius and should be given special recognition of his contributions to the program."

Fig. 10.1 Dick Carricato, electrical engineer.

"Ad Hoc" program area. Ad Hoc was the unclassified name for the Fulcrum study program for which I had been hired. Of course I didn't know that at the time. I was advised by one of these men that I should be very proud to have been selected because the interview-to-hire ratio for engineers was about 50 to 1.

Late in the afternoon I was assigned a desk in an area that I can only describe as a bullpen. There were only about six of us assigned to the area at the time, but there were many desks scattered about. One curiosity that caught my attention was a number of mechanical calculators, the likes of which I had never seen before. At first glance I was concerned that maybe these folks had not yet joined the digital computer age, but then reason prevailed as I realized that at least they weren't being used. I sat at my desk, and since I didn't have an assignment, my mind drifted back to the sequence of events that had brought me to this place.

Recruited by the Company

At my previous job at Norden, a division of United Aircraft, I had been consulting with Juan Rayces, a renowned optical engineer at Perkin-Elmer. Juan had engineered and overseen the manufacture of a special optical element that I had designed to be used in a head-up display system. That element had been delivered about a week earlier, so when my secretary told me some gentleman from Perkin-Elmer had called with an invitation to lunch I asked her to accept for me. I believed that this was going to be a technical marketing meeting to discuss how the optic had worked and also to determine how the company might help us in the future.

When I arrived at the restaurant I didn't see anyone I recognized. To my surprise, two men stood and waved me over to their table. They were Bob Landsman (Fig. 2.7) from the technical staff and a gentleman from their

Personnel department. They seemed to know me, and I wasn't terribly surprised that I didn't recognize them because I often had trouble remembering faces.

After introductions and the drink order, Bob looked at me and said, "Dick, we understand that you are unhappy with the project responsibilities at your present job. You would rather be working in a completely technical position in a more academic environment."

This just about floored me. These two strangers not only knew me by sight, but they knew things about me that I had never revealed to anyone at their company. Then, when Bob started asking technical questions, I started to realize that this was not the meeting I had anticipated. This was a job interview, and I wasn't looking for a new job. However, his opening statement had aroused my interest enough that I didn't end the interview immediately.

The interview was rather one-sided. Bob couldn't tell me anything about the job except that it would be technically challenging to a large group of physicists, mechanical engineers, and electronics engineers. I was able to gather a bit more detail from his questions, but never even a hint of what the job was about. I was quite impressed by Landsman's breadth and depth of engineering knowledge as demonstrated by his questions and the discussions those questions generated.

The questions concerning solid state devices and their applications were easy for me because that is what I did for a living. Further, I taught transistor courses in the evening graduate program at a state university. We then moved onto what was obviously Bob's main interest, servo (control) mechanisms and systems. I had never worked with servos; however, as an undergraduate I had been allowed to take a number of graduate-level courses, and one of these was a servo mechanisms course. While this didn't allow me to look like an expert in the field, it gave me enough knowledge to discuss the subject with Bob.

In conjunction with the servo discussions, Bob asked about my experiences with phase-locked loops and feedback amplifiers. My answer for many of these questions was that I had no experience with these things in the servo field; however, we use phase-locked loops, feedback amplifiers, and so forth, in the display field where I currently worked. I really wasn't concerned about my answers because I was not looking for a job.

When the discussion moved to servo and system computer simulations I started to get interested. A few years earlier I had required some computer programming for a project I was doing, so I asked our Data Processing department for help. The project couldn't afford the cost or the schedule quoted by our data processing people, so I purchased a copy of Daniel McCracken's book, *A Guide to FORTRAN Programming,* and taught myself

FORTRAN. Now I was involved with helicopter and submarine dynamic simulations, and I was anxious to continue that kind of work. Bob assured me that there would be plenty of servo and system simulation in the position they were attempting to fill.

I don't remember what we ate for lunch, nor do I remember how late we stayed, but I do remember that when I left the meeting I was hoping to hear from the company again. Obviously, I did hear from them again.

WORK BEGINS

When I joined Perkin-Elmer, I remember arriving at my desk in the tank, otherwise known as the mushroom patch, where all of the surrounding desks were occupied by people chatting and drinking coffee. The chatting stopped long enough for introductions all around and then it started up again.

There was a physicist from Eastman Kodak in Rochester who specialized in optical systems (a woman, rare in those days), another physicist from Itek with similar credentials, a servo engineer from Sperry, and an electrical engineer whose specialty seemed to be writing survey articles for trade magazines. The article he was working on at the time I met him was titled "Fuses, Everything You Ever Wanted to Know about Fuses."

I had been warned against speculating about the job we were all destined for, but I couldn't keep myself from trying to get some clues from the backgrounds of my fellow tanksters. I wasn't successful because the backgrounds were common to many others already there, and in the 1960s I didn't equate Itek and Kodak with space reconnaissance or spy satellites. For that matter, I doubt that I even knew what Itek did.

A short time later an engineer showed up in the tank looking for some help. I recognized him as one of the many people who had interviewed me after that first meeting with Bob Landsman, but I couldn't remember his name. He asked if anyone could simulate human response dynamics using an analog computer. Well, I had used an analog computer once in college so I raised my hand. What was it they told me in the Air Force? Never volunteer! That may have been true in the military, but here it was the way to get interesting work.

The problem was explained to me. It was supposed to be used for the lunar excursion module (LEM), but neither NASA nor the company making the LEM, Grumman, was ever mentioned. It was always the customer. I quickly learned that the customer is only called "the customer" when the customer doesn't want anyone to know who he is.

As I looked at the problem that I had volunteered to solve on an analog computer, it became obvious that I had no idea how to do it. However, it looked like the solution using a digital computer would be quite easy. So I said something like, "I can probably do this on a digital computer in less time than it will take to find the patch cords for an analog computer." My client, whose name I couldn't remember, was agreeable, and I worked late that night in order to have his answers in the morning.

The usual procedure for getting a program run on the mainframe computer was quite time-consuming. First, you had to write the program. Then, you submitted coding sheets to data processing staff who typed the information contained in the coding sheets onto machine-readable punch cards. If you were lucky the deck of punched cards would be available the next day, which you'd check for errors. If there were no errors (a rare occurrence) or only a few errors, you might be able to get your program run on the third day. Late on the third day, or perhaps early the fourth day, you could finally see the results of the program including any card punch errors you didn't catch as well as any programming errors.

This procedure had to be repeated whenever there were any errors in the program, so the programmer had to be ultra-careful to ensure that the submitted coding sheets were error free. A technique I used to speed things up was to do my own card punching. I wasn't as fast as a professional card punch operator, but I wasn't graded on how many stacks of cards I was able to punch in a day. The extra hour it took me to punch the cards accurately was small potatoes compared to the extra day to get punch errors corrected. I found that I could get about 200 lines of FORTRAN coded, checked, and debugged in 24 h if I stayed into the evening, did my own punching, and made friends with the computer operators so that my programs were run without delay.

The LEM simulations turned out well, and they led to more simulation tasks. The result was that I earned a reputation as a computer expert that followed me for the next 20 years.

The day that everyone in the tank awaited came for me on October 15, 1965. I was called to the Security office where I was given a bunch of documents to read and sign. I remember being awed by the words I was reading. It seemed like I was being told more than I needed to know. After 40 years the memory is a little hazy, but I do remember something like this: a study program leading to the design and development of a photo reconnaissance satellite, to conduct covert operations for the CIA undercover as the Discoverer program. This high-resolution system was to carry out search and surveillance missions over the Sino-Soviet Bloc. The program name was Fulcrum (the original name that was later changed to Hexagon).

The documents droned on about not revealing, acknowledging, or commenting on the existence of the program, the program name, the customer's name, or any of the participants in the program. This ban on discussion included everyone from one's family and friends all the way to others on the program with the proper security clearance but without an explicit need to know.

When I had finished all the reading and signing the security officer asked if I was surprised. I didn't have a feeling of surprise. I felt numb. I had just read a lot of words and concepts that I had never considered before. Covert operations, undercover, search and surveillance of the Sino-Soviet Bloc, and compartmentalized security clearances were all new phrases to me and quite foreign. I had a lot to learn! No, I didn't feel surprise; I felt like I had just joined the big league.

One distressing fact to come out was that this activity was only a study program that would hopefully lead to a proposal. That meant that there was a chance that we wouldn't win the job. What happens to all these new hires if that should happen? Did the company have enough work to support us all? This was a powerful incentive to work hard.

A photo-ID badge was prepared, and then I was escorted to the Ad Hoc program area where I was introduced to the security guard on duty. He took my other badge and gave me the photo-ID badge to use while in the classified area. To leave the area I would have to swap badges with the security guard on duty; the photo-ID remained under security control at all times. When visiting a secure facility for the first time, one needed a third party who knew both the visitor and the person being visited to formally introduce the two of them. These security procedures would become second nature after a while. This was standard operating procedure in the world of covert operations, the black world.

It turned out that I was not the only new kid in town. At about the time that I reported to the "Ad Hoc Suite," several others had also recently arrived. It was an opportune time for Bob Landsman to teach school. In truth, that is what he did much of every day during the first few months that I worked for him. Bob was the engineering leader of the program. His specialty was electrical engineering, but he was very knowledgeable in mechanical engineering too. He set aside about an hour and a half each day during lunch to describe the system as it was envisioned at that time. His office was a conference room with one wall covered by blackboards and a long conference table that provided adequate space to eat, take notes, and even ask questions when we weren't chewing. I never did learn when or if Bob ate lunch.

Bob's short description of the project was "an orbiting reconnaissance satellite system consisting of two independent stereoscopic cameras. We will be performing a search and surveillance mission from about one hundred miles in space. These cameras will be large. They will use up the space in an orbiting spacecraft that is about ten feet in diameter, about forty feet long, and almost 30,000 pounds." Now I was surprised!

The first U.S. orbiting satellite had been launched only seven years earlier, in 1958. Although we were now on our way to the Moon, the space program was still in its infancy. The Corona program, which was the first orbiting reconnaissance satellite, began attempting launches in 1959. Like a true pioneer, that program broke the ground, endured the failures, and learned many of the lessons that would be so successfully applied by Hexagon. In 1960 Corona began having successful flights. It continued in operation until 1971 when its mission was taken over by Big Bird, the name given to Hexagon by the media people who became aware of a new larger satellite being launched from Vandenberg Air Force Base. Information about Corona was not part of Bob's lectures because most of us were not yet cleared for that program. One interesting fact that we learned about a year later was that the film supply carried by Corona weighed 20 pounds on the early flights. That film load grew to 180 pounds as Corona evolved and matured. Hexagon would carry many times that amount of film, enough to stretch from our building in Connecticut to New York City. The film would be returned in four recovery vehicles that would be caught in midair while descending by parachute. And again I was surprised! There would be many more surprises before these luncheon lectures were finished.

The coverage from 100 miles altitude is approximately 32 million square nautical miles. This coverage is with full stereo and does not include the redundant coverage of area caused by additional frame overlap at wide scan angles. The total land area of Earth is about 40 million square nautical miles.

There were supposed to be six launches per year, and the cameras were to be capable of operation for 45 days. That would mean that the photo analysts would have 15 acres of film to look at using high-powered microscopes every year. Luckily that never happened because the analysts would have been overwhelmed. These requirements were generated based on what the current systems of the early 1960s could do while allowing for expected improvements in the technology. However, Hexagon's reliability was so good and its performance was so high that we never had to fly more than three missions in one year, and most years we only flew one mission with the cameras remaining operational almost all that time.

Servo-control System Basics

The camera system consisted of two independent cameras. This meant that if either camera should fail, the other could continue the mission. Of course, this would mean that the information gained from stereoscopic viewing of the imagery would be lost, but the mission would not.

The rotation rate of the optical bars was controlled so that adjacent swaths overlapped slightly assuring that there were no missed areas in the photography. The +/− 10-degree pitch angles provided a 20-degree stereo angle for stereoscopic viewing of the imagery. An advantage to this stereo angle was that it placed a target viewed by the aft-looking camera three-and-a-half frames after it was viewed by the forward-looking camera. If a target was at the edge of the film on one camera (the camera's lowest performance area) it would be in the center of the film on the other camera (the area of highest performance).

The maximum *V/h* specified by the CIA was to be 0.054 radians per second. This set the rotational rate of the optical bar as well as the linear speed of the film. The best resolution was at *perigee*, where the vehicle was closest to Earth. That was where the rotational rate of the optical bar and the linear film speed was the maximum, approximately 30 revolutions per minute of the optical bar and 204 inches per second across the slit at the focal plane.

Important to note is that the image formed by the optics rotated as the optical bar rotated, scanning the scene below. To have the motion of the film match the motion of the image therefore required that the speed of the film must match that of the image very precisely, and the angle of the film must track the angle of the optical bar during photography.

My experience with photography prior to joining Perkin-Elmer was taking family snapshots. It was pretty easy, just get everyone to "hold it" and snap. We tell everyone to hold it because the film in the camera is standing still, and we would like the image and the film to be going at the same speed so there will be no smearing of the image on the film. Well, this seemed almost impossible to accomplish: The spacecraft is moving over the ground at just over 4 nautical miles per second in its flight direction. The scanning mentioned earlier would make the ground look like it is traveling an additional 250 nautical miles per second across the flight path. Because there is no way to hold the scene of the ground still, we had to move the film so that it would be in almost perfect registration with that moving image!

My engineering instincts kicked in, and I asked at one of Bob's lunches just how perfect was almost perfect. Bob explained that the optical bars would be rotating at a maximum speed of 3.4 radians per second or about 195 degrees per second. The focal length of the optics was 60 inches, so the image velocity at the focal plane was about 204 inches per second. If we could keep the error less than 0.02 inches per second, that should be perfect enough. I think my voice squeaked a bit when I reminded Bob that this was less than one one-hundredth of one percent. This number went into a special place in my notebook. This would certainly be one of the many major challenges on this program.

Bob casually mentioned at this point in the discussion that there would be many sources of error trying to defeat our attempt to realize this small error number. These would include bearing noises, roller nonuniformities, other torque perturbations, electronic noise, and computational errors. From the early days of photography, engineers had used devices to make the film transport easier to design. Among the list are gears to accommodate typical motors, sprockets to help guide the film, and flanged rollers to keep the film from running off the rollers. Each of these is a known source of film tension perturbation, which would lead to velocity errors. So a rule was adopted that none of these devices would be allowed in the design of the Hexagon system. The film would travel from the supply, through the camera and to the take-up, a distance of hundreds of feet and over hundreds of rollers without help from these error-inducing devices. In their place a film steering device would have to be invented.

These decisions made the life of the mechanical engineers challenging, to say the least. They also led to a system that performed better than anyone had dreamed possible.

For many reasons discussed in Chapter 11 it was decided that all motors in the film path would be DC motors. However, space flights at that time were experiencing problems with the brushes that were needed to carry electric current to the motor's windings. Most of the problems were caused by the materials used in the brushes operating in the vacuum of space. So it was decided that all of the motors would be brushless! There was one problem with this requirement: brushless motors did not yet exist. We would have to develop our own.

Further, and for the same reasons, all velocity and position sensors would also be brushless, either optical or magnetic. As the design progressed, optical encoders were chosen for use as servo tachometers, position sensors, and function generators for the trigonometric functions of the optical bar scan angle that would be required for image motion compensation (IMC).

While the optical bar is rotating through 360 degrees every 2 s, pictures are recorded only during that 120-degree arc which is +/− 60 degrees on either side of nadir. That means that the film is at 204 inches per second for one third of the time, and the average film speed is only 70 inches per second. To make use of this fact requires a storage device that will allow the supply to input film at 70 inches per second, the take-up to remove film at 68 inches per second, and film at the focal plane to move at speeds of 204 inches per second (provided that the average speed is held to 70 inches per second).

The device to accomplish this was known as the *looper*. A detailed explanation of the film path in the looper and of the precision film velocity control is discussed in Chapter 7.

The looper allowed the coarse film path to run at the average speed of the film being used in the fine film path. It also made possible the recycling of the film so that none was wasted between frames. The invention that made the entire concept possible was invented by the late Don Cowles, a mechanical engineer, and patented by Perkin-Elmer (see Appendix B). This relatively simple device, consisting of a few rollers and two pivoted air bars, was called the *twister*. The twister was a self-aligning, passive device that allowed the film to be rotated in synchronization with the optical bar during photography. It was capable of rotation through 180 degrees; therefore, the angular velocity profile for the film looked similar to the linear velocity profile of the film. Although the film's linear velocity profile changed with the various scan modes, the angular velocity profile remained constant.

The *platen* was the device that physically contained the twister, and actually rotated the film in sync with the optical bar. Among the many precision mechanisms contained in the platen were

- The metering capstan (the servo control system) that maintained the required precise matching speed and position between the film and image along the length of the film.
- The fail safe shutter and its controls that guaranteed an electronic failure in the shutter would not totally disable a camera.
- The slit that was the exposure control for the camera.
- A focus and tilt mechanism to assure proper focus across the entire film width.
- The platen-to-optical-bar servo sensors that acted as the position and velocity sensors providing feedback for the control of the platen during photography.
- Flex cabling to carry the many signals and power leads needed in the platen.
- Flexible light pipes to carry the film-edge marking data.

Some system design rules were arrived at during the study phase.

1. In order to increase reliability of the system not only would the motors be brushless, but all of the tachometers and position sensors would be brushless. That is, they would be either optical or magnetic. (We were 99-precent sure we wanted to go optical, but the ease of manufacture of magnetic devices kept the option open for a while longer.)
2. To reduce disturbances to the space vehicle, all large inertia devices such as the supplies, take-ups, platens, and optical bars would counter-rotate in the two cameras in order to minimize momentum unbalance. They would maintain constant velocity during photographic periods.
3. The supply and take-up servos of each camera counter would rotate to reduce some of the uncompensated momentum associated with single-camera operations even if one camera was not in operation.

4. All velocity servos would have a position loop to drive the velocity error to zero. An accurate feed-forward velocity command would be employed, and where necessary, a feed-forward torque command would be used.
5. For reliability reasons, no potentiometers would be used. The ganged potentiometers in the supply and take-up adaptive servos and the tension sensor position transducers were the exceptions to the rule. Rigorous reliability analyses of competing devices resulted in the selection of pots for these applications.
6. No electromechanical relays! This rule had to be broken when we learned that the command interface from the spacecraft depended on dozens of latching relays on our side of the interface.
7. To further reduce torque disturbances on the servos, the brushless motors would be non-cogging devices.
8. As the design progressed many more design rules would be adopted.

Chapter 11 discusses how the servo-control system was designed and built.

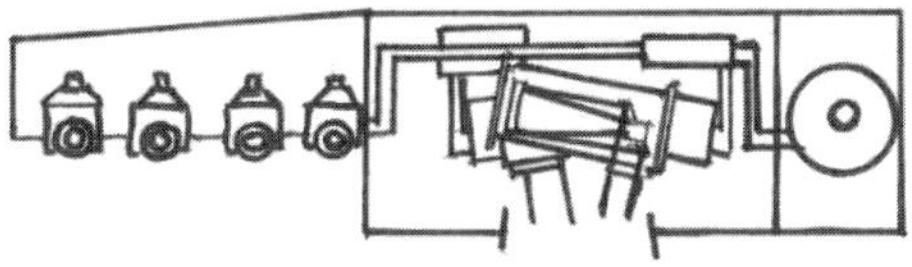

Chapter 11

SERVO SYSTEM

DICK CARRICATO, MARTY YELLIN, AND RICHARD LABINGER*

FILM PATH CONTROL SYSTEM

Hexagon's film path control systems were required to control film speed, film position, and film tension to accuracies that had never been achieved before. In order to accomplish this, we very quickly discovered that we had to either use state-of-the-art components or invent them. In addition, because ours was a long-life space application, it required outstanding reliability, in many cases equal or better reliability than necessary for manned space applications. This led to many new implementation developments.

The film path was designed using complex feedback control systems (called *servomechanisms*) to precisely control to state-of-the-art accuracies, the film's velocity, position, tension, and synchronization to the rotating and translating image.

The key feature of a servomechanism is that it is an electromechanical feedback system used to control a desired output state (position or velocity or both) with extreme accuracy. The key word is *feedback*. Instead of controlling the specific position or velocity of a mechanism by open loop commands, we instead controlled the error between the desired position and velocity as well as the actual position and velocity. This error is created by sensing the output state (position and/or velocity) and subtracting it from the commanded state. That error is used as the signal to command the mechanism motors. By driving the error towards zero we achieved our desired goals. There are three key interrelated design considerations that the designer needs to take into account in order for the servomechanism to achieve its desired goals: bandwidth, accuracy, and stability.

Copyright © 2013 by the American Institute of Aeronautics and Astronautics, Inc., All rights reserved.

*In the first part of this chapter, the authors explain how the system worked. The second part contains the recollections of Marty Yellin and Richard Labinger, electrical engineers whose careers at Perkin-Elmer coincided from beginning to end with mine. They were both experts in servomechanisms. Marty was the leader of the group of servo engineers and was usually the one who presented the servo analyses for each of the major subassemblies at the design reviews with our customer.

Bandwidth refers to how fast the servomechanism can respond to a disturbance. As the disturbance frequencies get higher, the servomechanisms bandwidth must also get higher. This is exactly analogous to the bandwidth limitations of an audio amplifier that prevents us from hearing certain low and high frequencies from the music source. Mechanical bandwidth limitations are driven mainly by mechanical loads; that is, inertia, stiffness, and friction. Electrical bandwidths are limited by the electronics.

Accuracy refers to the allowable position or velocity error desired when confronted with various position and velocity commands such as step functions, ramp functions, trigonometric functions, etc. The servo loop also needs to attenuate disturbances that could affect accuracy. These disturbances could be bearing noise, electrical noise, environmental load disturbances (like the wind gusts on a telescope), etc. Estimates of these disturbances are made with the help of electrical and mechanical experiments, analyses, and experience.

Stability is dependent on servo bandwidth. As the bandwidth increases, mechanical and electrical breakpoints occur that cause phase shifts between the output and input signals. As the phase shifts increase towards 180 degrees, the servo system will become uncontrollable and unstable unless appropriate design compensations are incorporated.

A major subassembly of the system, the film drive, had a roller/motor called the film drive capstan that pulled the film along. The film drive was attached to the rear of the platen. As the optical bar rotated, it successively imaged narrow slits of ground imagery onto the synchronously moving film. During one frame of exposure, a swath of the ground about 10 nautical miles wide in the in-track direction by 170 nautical miles in the cross-track direction (at 100 nautical miles altitude) was captured. If the film was kept moving during the nonimaging portion of the optical bar rotation, there would be an unacceptable waste of film. In fact, if we simply ran the film at constant velocity we would waste between 83 and 96 percent of the film. Similarly, we could not start and stop the film during the optical bar nonphotographic periods because this would require infinite (or at least very high) accelerations and decelerations of the film spools. This is not practical and would require a great deal of power.

A walk around the film path transport system showing all the servomechanisms is shown in Fig.11.1. The job of all the servomechanisms in the film transport system is easily stated: get the film off of the supply reel at a constant and relatively slow speed, pass it through the focal plane at as high velocity (in rotation and translation) as the image on the ground, and then place the film on the take-up reels again at a constant and relatively slow speed. At the same time, the servomechanism must control the film flatness

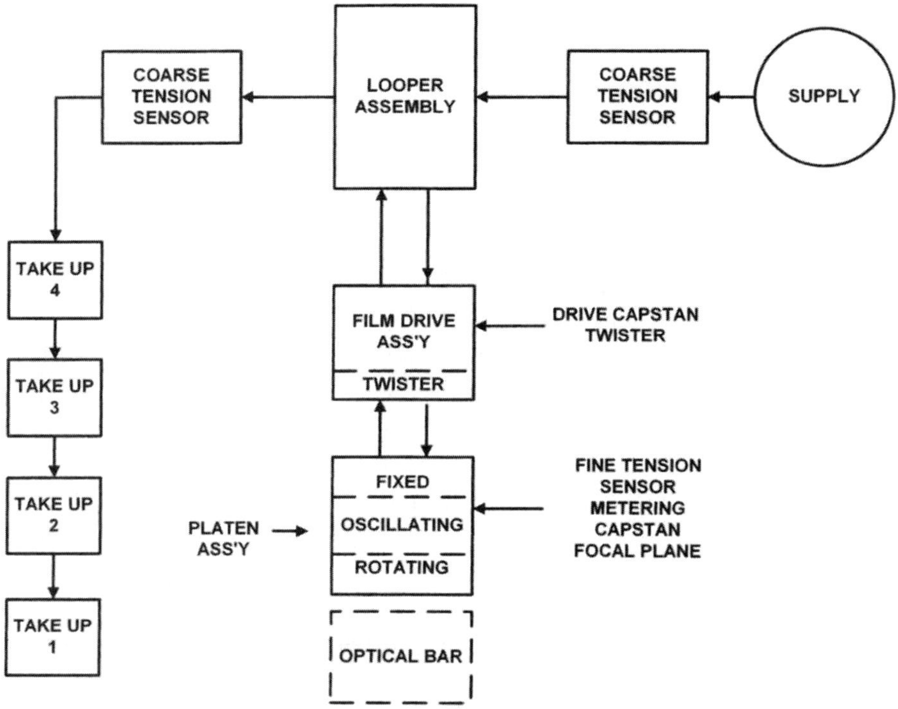

Fig. 11.1 Film path.

and reverse the film path during nonoptical imaging times so that no film is wasted. This may sound straightforward, but it was quite difficult to achieve.

The actual film velocity profile is shown in Fig. 11.2. The film is accelerated until it reaches exposure velocity and then it is advanced at

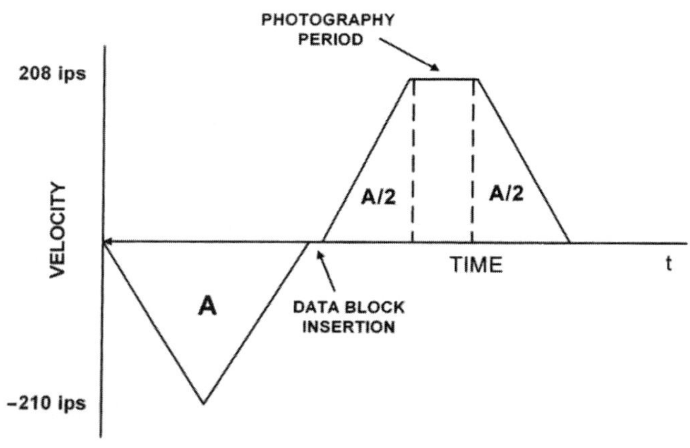

Fig. 11.2 Fine film path velocity.

constant velocity during photography. At the end of the exposure time, the film is decelerated. As it slows down, it reverses direction and eventually stops for a short time so that a data block mark could be imaged onto the film for identification purposes. This process is then repeated for the next frame. In this way, images are placed on the film with only 2 inches between them and minimal wastage of film.

There were two optical bars continuously rotating in opposite directions to cancel out angular momentum. They each performed a whisk-broom scan of Earth as the satellite moved forward. Although the optical bars rotated continuously, photographic imaging occurred only during +/−60 degrees or less of rotation depending upon orbit altitude and the desires of what images were required (wide angle or close-up imagery) and where the images should be with respect to nadir. It is during that period when the film's linear velocity and rotational speed in the platen have to be synchronized exactly with the image at the optical bar focal plane. Image motion compensation (IMC) must also be performed so that there will be no significant residual image smear caused by the rotation of Earth during a photographic exposure. IMC is discussed in Appendix C.

During the remaining 240 degrees or more of nonphotography rotation of the optical bar the film path must decelerate, reverse its direction, and back up in position. Then it must accelerate up to speed just in time to begin another imaging cycle (Fig. 11.2). The profiles to perform these maneuvers are generated from synchronizing signals on the optical bar encoders. The same encoder also contained additional tracks that generate the IMC profiles that drive the metering capstan and platen servos.

The optical bar servos operated at constant velocity with zero positional error so that ground imaging would always take place at the desired optical bar angle. The optical bar servo's summed error signal served as a useful diagnostic in that the magnitude and frequency of the observed error provided insight on the status of bearings and optical bar pressure seal torques. In fact for the prototype system, it indicated that the optical bar seal friction torque was excessive and had a once-per-revolution high point.

The film had to leave the supply reel at constant velocity. An accurate velocity servo is usually not difficult to implement. However, as film is spooled off, the supply reel's mass and radius is constantly changing. This results in the moment of inertia changing by the fourth power of radius. Since moment of inertia is a key contributor, the servo has to be adaptive to accommodate these dynamically changing parameters.

Because the radius was being reduced, a method was needed to speed up the angular velocity of the supply motor in order to keep the film velocity constant independent of radius (film velocity is the product of angular velocity multiplied by radius). In order to sense the change in velocity of the film coming off the supply, we used a device called the *coarse tension sensor*.

The coarse tension sensor consisted of three rollers in a Y configuration with the bottom roller connected to ground via a negator spring (a constant force spring). Any difference in velocity between the supply input and the output roller caused a displacement of the bottom roller up or down. A transducer (potentiometer) coupled to that roller sensed this displacement and sent a correction signal to speed up or slow down the supply motor to keep the velocity into and out of the coarse tension sensor equal. In 1965 through the early 1970s all this had to be done using hardwired analog computation circuits. There were no microprocessors.

After the film left the supply, it had to be moved in accordance with a prescribed film velocity profile to enable photography to occur at the proper time and to utilize as much of the film as possible (see Fig. 11.2). The film was accelerated to photographic speed, during which time the exposure occurred. It was then decelerated and driven backwards so that the next exposure was made with only two inches of film between exposures. The film was then stopped so that a data block could be placed on the film in this space. The job of accommodating the film profile from constant velocity at the supply to variable at the focal plane in the platen was accomplished by means of a film storage device called the *looper*. Directional changes were achieved by use of devices called twisters and air bars discussed in Chapter 9.

The Hexagon system was designed and built using the most stringent reliability requirements of any spaceborne program in existence at the time. This dictated the use of solid-state components, but only those with an established track record for reliability. In the 1960s and 1970s, analytic functions could not be configured practically using digital hardware or software because there were no microprocessors and memory was prohibitively expensive and large. Therefore, analytic functions such as film velocity profiles and electronic circuits such as servo error detectors, function generators, and electronic amplifiers had to be designed using analog circuits. Analog circuitry proved to be a great challenge.

We were allowed to use some newly developed solid-state operational amplifiers instead of vacuum tube amplifiers. These were among the first integrated circuits ever built and were really hybrid circuits in that they were comprised of discrete components (transistors, resistors, capacitors, etc.) encapsulated in a can. They were smaller and lighter than discrete component amplifiers laid out on a printed circuit board would have been. However, as good as they were for Hexagon applications, we had to accommodate their idiosyncrasies, such as thermal drifting, voltage offsets, device-to-device differences, and gain variations. The electronics are described further in Chapter 17.

The motors required to drive the optical bars, supply reels, take-up reels, platen, and capstans had to be compact, lightweight, and reliable. The motors had to be direct drive so that they could be mounted directly on the shafts that they drove. A direct-drive motor provides the highest practical torque to inertia ratio where it counts—at the load.

Motor commutation is traditionally performed by the use of contacting brushes. They feed current to the coils, as required. We did not want to use brushes because contacting brushes inherently have limited life and as such will wear down and create contamination. Such contamination could drift in a weightless environment and thus redeposit in critical areas, such as in bearings or on optics. This could not be tolerated. Secondly, contacting brushes generate electromechanical interference (EMI), or noise. This could be disastrous to our sensitive electronic circuits. Instead motor commutation for our DC motors was accomplished by using one of the tracks on the optical encoder that was attached to each motor shaft. This track had alternating opaque and transparent segments that corresponded to the commutation positions required. So, as the shaft turned, current was switched to the appropriate coils by power transistors driven by the switching signals derived from the encoder.

Therefore direct-drive brushless motors were ideally suited for our high acceleration applications with numerous rapid starts and stops. In order to save power we also wanted a high torque-to-power ratio. The more poles and volume of copper in the windings, the higher the torque. We looked for a supplier to help us with this component. Hank McClansky owned a very small motor manufacturing company named Mac Bar in Huntington, Long Island. He came up with a method of putting up to 30 percent more copper windings within a given volume. He did this with a unique winding system. His expert winder was his wife, and they worked out of their kitchen and garage. For a given horsepower requirement his motors were lighter and required less power than any other motor manufactured in the United States at that time. He was a regular supplier of motors to the military.

He had also worked with Perkin-Elmer before this program and during its proposal phase. For Hexagon, electrical engineers Gene Aha and Richard Fink defined the motor requirements and monitored Mac Bar. McClansky made many motors for the program, but when the production quantities and schedule pressures increased, he could not keep up with the work. Perkin-Elmer decided to bring some of the motor work in-house. An unclassified production facility was set up, and both our company and Mac Bar fabricated motors that fulfilled the needs of the program. The motors were reliable and there were no failures.

As previously stated, motor commutation was accomplished by use of optical encoders. In addition, we used optical encoders to act as digital tachometers for velocity feedback and to generate digital signals of various analytic functions such as IMC. The encoders generated incremental signals or pulses as the encoders turned. Tracks used for servo velocity feedback had equally spaced marks generated at constant frequency if the shaft velocity was constant. Other tracks had marks spaced in accordance with a desired analytic function. The tracks were all imprinted at various diameters of the encoder's glass disc.

The development of optical encoders was facilitated later by the invention of the light-emitting diode (LED), solid-state photo-transistors, and photo diodes. However, at the beginning of the program, LEDs were too new and therefore had no track record, so we had to use miniature incandescent lamps. These were only designed for commercial use and were extremely variable in performance. We had to buy ten to one hundred times the quantities we needed and then select the best ones. Each dollar bulb ended up costing us hundreds of dollars, as discussed in Chapter 17. It was much later in the program that we were finally allowed to use LEDs. The person responsible for the system encoders was Peter Van Raalte.

The servomechanisms designed and fabricated for the Hexagon program represented a challenge to the design engineers that went beyond anything that had ever been done before. Considering that this work was done prior to the existence of personal computers, digital calculators, and analytical software, it is a testimony to the outstanding skills and abilities of all of those dedicated engineers who worked on it. It was an amazing accomplishment that culminated in the fact that every servo worked to perfection during every mission without a failure. The Hexagon program, and indeed the nation, owes a great deal of gratitude to those who worked so diligently to achieve this tremendous result.

Marty Yellin's Story

On June 15, 1965 I interviewed at the Perkin-Elmer Corporation. I was 26 years old with a Masters Degree in electrical engineering. I had recently enrolled at NYU in a new doctoral program in biomedical engineering; I was interested in joining the company because they were involved in automated techniques to detect blood cancers and chromosome defects using advanced pattern recognition techniques and adaptive control systems. I was interviewed by about half a dozen brilliant people and left at the end of the day feeling I was too incompetent to be hired there. Amazingly I was hired and soon was deeply and happily involved with the mitotic cell scanner and other biological pattern recognition projects.

I was totally unaware that while I was doing this work, my special program clearance process was going on. About 6 months after I was hired, I was told that I was being transferred across the street to head up a control systems group on a large program. I told my boss I didn't want to leave the projects I was involved with. He told me not to worry and assured me that I would only be gone for a short time. That short time turned into about 7 years of the most intensive and rewarding work I ever did in my 40-year career (Fig. 11.3).

Into the chaos. The first stop on my journey was to the Security office to be briefed for this program by Bob Markin of the Security department. He couldn't tell me who the customer was. I couldn't understand why I wasn't allowed to know this information. I remember thinking that maybe we were doing work for China or Russia! I really believed that.

I moved to the building where the program was being implemented. It was utter chaos. There were hundreds of people all packed together in a large room without offices or partitions. Boxes, books, drafting tables, monstrous mechanical calculators, and slide rules were everywhere. The room was divided by discipline, so that in each area the discussions and technical issues were different. It was like a Tower of Babel! As I headed towards my area I saw that I was going to be heading up a group of about 15 engineers, most of whom were much older than I.

The briefings ran from 8 a.m. to about 8 p.m. every day for the next 5 days. The enormous challenges of designing the servo systems started to become apparent. For one, almost every servomechanism was coupled by film that acted as a "wet noodle" in the lateral direction and as a nonlinear spring in the linear direction. The one servo not coupled by film was the optical bar servo, but it had to be electronically synchronized dynamically to the metering capstan and platen servos compounded in difficulty by the addition of IMC in both the rotating and linear directions. In addition, there were rotating assemblies such as the supply and take-up reels with inertia that changed substantially as the film wound off the supply and onto the take-ups. This

Fig. 11.3 Marty Yellin, electrical engineer.

required a level of sophisticated adaptive control to ensure that the servo loop remained stable.

The bottom line was that we had to design about 10 servomechanisms that were all coupled nonlinearly in order to ensure that the film speed, film position, film tension, film lateral control, image synchronization, IMC, and other requirements all met extremely accurate requirements. To do all this without any access to computers was an almost impossible task, and I now had to come up with a detailed plan including schedule and costs and present it to my boss in 1 week! Somehow it all came together, predominantly because of the outstanding people who worked for me and with me.

I thought this is an incredibly complicated system. It turned out that we invented as we went along. Money was no object. Schedule was everything. It was an overwhelming task but was exhilarating work. We had the proper motivation, a common objective, and teamwork, as well as management that made us feel like we could do anything.

My team was the best thing I had in heading up the servomechanism group. The engineers working for me were brilliant. All the servomechanism analysis and design was handled by just 15 people who worked around the clock with me, practically 7 days a week. There were many good people, but a few stand out as exceptional. Richard Labinger comes to mind. He was responsible for the servo electronics. He had to be responsive to constantly changing requirements and electronics technology as well as the ongoing reliability, packaging, thermal, and power constraints. Coupled with all of this were the constraints of meeting critical deadlines and staying within budget. The program could never have succeeded without him.

Another key team member was Howard Burg. Howie was responsible for building working brassboards of each critical servomechanism. You must remember that there were no convenient computers available at Perkin-Elmer for monitoring such complex servos. Our approach instead was to build working models of the servos as we were designing them. In that way, we learned about real problems that needed solving. Howie was instrumental in building each servo and later in building and operating an engineering model of the whole film path.

There was no way that I could account for the "wet noodle" problem within my group. Thankfully a team of system engineers headed up by Franz Holzer undertook the development of an end-to-end digital model of the entire film path. Key inputs to this model came from Bill Moore, Steve Dubowsky, Dick Carricato, Bob Landsman, and Karl Herring. Without their outstanding work I never could have verified that the servomechanism designs were adequate using uncoupled analyses and tests. This turned out to be extremely important during the design reviews.

Our customer had hired a large aerospace company (TRW Corporation) to oversee the servo design. This company was known as SETS (System Engineering Technical Support) but I called them Zetz because they were always trying to show how smart they were at my expense, especially in front of the customer. Because they were a large company with many resources, they had developed very sophisticated digital models of the film path that they used to try to convince me and the customer that our designs were inadequate. Thanks to the systems models developed by Holzer and others, we were able to prove that our designs were valid. Even so, later I became good friends with many of the Zetz personnel.

Very shortly after I joined the effort, design reviews began. These reviews were presented for each mechanical assembly (supply, capstans, optical bar, platen, take-up, etc). Because each of these assemblies contained a servomechanism I was called upon to give an overview at every preliminary design review (PDR) and customer design review (CDR). In fact I gave presentations 16 times using over 400 view graphs that I think, if not a record, came pretty close. The presentations were held in a room called the executive conference room. In those days almost everybody smoked. Standing in front of the room all one could see was a smoky haze rising above the 50 to 100 people attending the review. In the front sat the customers, Zetz personnel, and key Perkin-Elmer management. Further back were the engineers and designers. Toward the back were the perpetual meeting goers. These were people that attended every presentation, but I never was able to figure out what else they did.

At each design review I explained the detailed servo requirements, presented the design at that point in time, and shared any analytical or test results achieved so far to verify our design approach. There was a lot of pressure on me; I was responsible for a lot of difficult design areas. In the beginning I was very nervous but as the months and years went by I became very confident, and indeed somewhat arrogant in my presentations. I am reminded of a few examples of that arrogance: once while justifying a difficult design approach, I was asked a somewhat doubting question about my design from one of the Zetz personnel. I answered sarcastically, "If you have to ask that question, then you will never understand my answer!" Another time after a long tirade on how the platen servo worked, I asked "Any questions?" to which one of the customer personnel answered, "who would have the guts to ask you anything!" Another time I started my presentation by saying "My name in Marty Yellin. Any questions so far?" Looking back, I am sorry that I was so arrogant. With the benefit of hindsight I see now that it was really a defense mechanism for my own feelings of inadequacy about meeting all the difficult objectives I was responsible for.

Richard Labinger's Story

April 19, 1965, a holiday called Patriot's Day in Massachusetts, enabled me to be interviewed for a position at Perkin-Elmer without having to take a day off from my job at Raytheon. The interviewers were Bob Landsman (Fig. 2.7) and Dick Carricato (Fig. 10.1). They proceeded to describe, in exotic detail, the state-of-the-art technical challenges I could work on, but were so vague on the application that I left envisioning the most unimaginable scenarios. I couldn't tell if they were crazy, were hustling me, or were offering me the career opportunity of a lifetime. Needless to say, I decided to find out, and so reported to work as an electrical engineer on July 5, 1965. For 3 months I was kept busy doing nothing. The highlight of this period was trying to find automobile antifreeze in the middle of summer to test a stratoscope assembly—not an easy task.

When I finally received my clearance in October, 1965, I was allowed into the newly acquired factory across the street (Fig. 11.4). This building had been hastily converted into working space to be used by engineers, draftsmen, and lab technicians. Our desks were in a big, open room that seemed to have little or no climatic control, but had ankle-bruising electrical outlets sticking up everywhere. Mike Maguire (Fig. 2.7) and his cadre of directors had paneled offices. In response to our complaints, he said "the factory stops here." The room had no windows and there were armed guards at every entrance. The windows that did exist were all covered up.

Fig. 11.4 Richard Labinger, electrical engineer.

I was exposed to specifications and concepts for a spaceborne reconnaissance camera system more bizarre than anything I had imagined. But at least I was now on the team! We were going to do things that had never been done before. We were going to push the state of the art in every electromechanical/optical discipline required for photographic imaging, all for a spaceborne application. I had to learn the meaning of terms I had never heard before, such as IMC, "V over H" (for *V/h*), loopers, air bars, and twisters to name a few. Because my area of expertise, such as it was when I was 30, was servomechanisms, my first assignments pertained to electronics for the control of the film transport system.

Our first major objective during the months in this facility was the design and construction of a brassboard of the optical bar servo and the film transport system. As soon as designs were laid out on paper, technicians grabbed them and began building circuits. At this stage there was no time for mundane tasks like testing new designs. Everything we did was geared towards getting that brassboard ready to show to the customers (whoever they were).

I think we spent about 2 years there before moving to the newly constructed facility in Danbury. Working conditions were considerably better. We had a bright new office area with four-man cubicles. Considering that we spent 50 to 60 hours a week there, the new working conditions were appreciated.

I first worked for Bob Landsman, then Bob Williamson. I worked with Marty Yellin (Fig. 11.3) on the optical bar servo and the synchronization between the metering capstan and the platen. Stan Anhouse and Gerry Leblanc worked on the sequencer.

It seemed to me that during the first couple of years our most important product was the view graph. We were constantly preparing customer presentations. There were monthly briefings, PDRs, and CDRs. And, of course, prior to every presentation we had to conduct dry runs and brief management.

At my level in the corporate hierarchy at the time, I wasn't privy to knowledge of the identity of some of the customer personnel who attended these briefings. I gave a briefing one day on a multibit phase detector I had developed for the phase lock loop to control IMC. It was a major achievement. I designed it and made it work, but I couldn't explain it. I was peppered with questions and comments from an individual I had never seen before. He asked insightful, probing technical questions that I had difficulty answering. He then sarcastically answered some of them himself. Needless to say I was devastated and quite shaken. It turned out that he was Dr. Richard Garwin, one of the highest level technical advisors in the government at the time. Afterward, Mike Maguire consoled me by telling me

I was in good company. It turned out that Dick Garwin had taken smarter people than me apart. He took everybody apart. Rod Scott, our chief scientist, came to see me a day or two later and he took all my notes and drawings from my notebook and he went off by himself and he eventually was able to explain how this thing worked.

We were all working excessively long hours, which made family life tough. During Thanksgiving in 1968 most of us were given a 4-day weekend. However, we had a major test underway that was not going well, so some of us came in to work. I remember being in the clean room taking test data on Thanksgiving night when behind me I heard "how's it going?" It was Mike Maguire. He knew some of us had given up our holiday weekend, but he came in to let us know he was aware of our efforts. This meant a lot to us and was one of the reasons we thought so highly of him. Another was his way with words. One time one of the guys went into Mike Maguire's office with a box of cigars. He said, "Mike, my wife just had our sixth child." Mike answered, "Congratulations, you have half a family." Mike had 12 kids.

Among the most difficult challenges in the Hexagon design were the changes. Everything was evolving at once. We no sooner completed and released a design to production when we were handed a new dynamic property of the film path that had to be accommodated. This impacted schedules. Eventually, everything came together and we completed the assembly and testing of the systems and they were shipped, integrated with the launch vehicles, and launched.

To the best of my knowledge, the film transport servos performed as designed and contributed to successful missions. It was a career opportunity of a lifetime. We felt we were doing a service for the United States, doing intelligence, not weapons. The Perkin-Elmer culture was very structured and disciplined, and this was important to our success. When the program was declassified in 2011, a big burden was lifted because we could then talk about what we did and that the job was very substantial and never done before.

In the next chapter, the frame and take-ups are discussed.

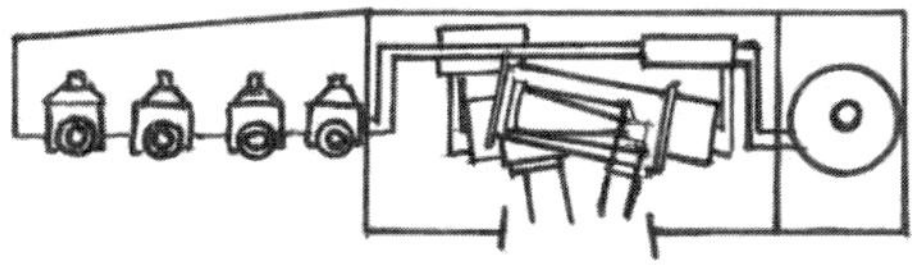

Chapter 12

FRAME AND TAKE-UPS

DON FISH, TED URBAN, AND PHIL PRESSEL

Neil Freeman was the engineer responsible for the design and fabrication of the frame. He was kind and modest; he was my friend, but unfortunately Neil died several years ago. I am writing the frame portion of this chapter on his behalf to acknowledge his fine work. Neil walked with the help of a cane having had polio in his youth. He had a cute habit of tucking the cane's handle in his back pocket when standing talking with someone.

THE FRAME

The frame was the structural support pallet or backbone for most of the components of the Hexagon system. It supported the following major subassemblies on separate mounting pads:

- Two optical bars
- Two platens (mounted to the optical bars)
- Two film drives (mounted to the back of the platens)
- Two loopers
- Two cross-over assemblies
- Nitrogen pressure tanks
- Parts of the film path articulators and cross-overs
- Some of the electronic boxes
- S-cubed (cameras added later during the program)

The mounting interfaces on the frame for the two optical bars positioned them at their required 20-degree stereo angle, namely at +10 degrees for the A-side optical bar facing forward, and at −10 degrees for the B-side optical bar facing aft.

The name of the assembly that comprised the frame, the optical bars, and the other items stated above was called the two-camera assembly or TCA for short. The TCA was located in the midsection of the vehicle.

Copyright © 2013 by the American Institute of Aeronautics and Astronautics, Inc., All rights reserved.

The frame did not support the film supply and take-up assemblies. The supply assemblies were in the aft portion of the vehicle and the four take-ups were located in the forward section of the vehicle.

The frame was made of long lengths of aluminum sheets 0.125 inches thick, bent and formed into a box-type structure, then riveted together. There was also a layer of epoxy between the sheets in order to provide extra holding strength. A wooden model of the frame is shown in Fig. 12.1.

Optical alignment cubes with mirrored faces perpendicular to each other were also mounted to the frame to provide an accurate optical reference system.

The frame itself was mounted to the ten-foot diameter vehicle midsection by two fixed pads at the aft end of the vehicle and by two vertical linkages at the forward end. The two links had spherical bearings at each end. The purpose of the vertical links was to provide strain-free motion of the frame relative to the vehicle during the launch. This prevented high redundant loads from passing through the frame and prevent permanent deformation of critical aligned surfaces.

Highly detailed models of the TCA, and eventually of the TCA and vehicle assembly, were created in a computer modeling technique called NASTRAN. Modern versions of NASTRAN are still used today when dealing with structural engineering. These models were used to predict resonant frequencies and stresses of critical assemblies.

Vent holes were provided at numerous places in the frame to permit venting of trapped air inside the hollow frame during launch. Once in space the interior space of the frame was at the same pressure, namely zero, as the exterior. Filters covered all of the vent holes in order to prevent any contamination debris from entering or exiting the frame space. This was one

Fig. 12.1 Wooden model of the frame.

of the precautions taken to make sure that no contamination debris would float around in a zero *g* environment and deposit itself somewhere unwanted, such as onto one of the optical elements.

Because the temperature environment of the TCA had to be controlled, it was necessary to control the emittance of the large surface areas of the frame to a value less than 0.10. A low emittance, or emissivity, is required to reflect heat away from the TCA. The heat source could be the sun, electronic boxes, or even albedo (reflected heat from Earth, about 15 percent of the sunlight). A low emittance meant that all of the exterior surfaces of the frame had to be buffed and polished to a highly reflective finish.

In 1970 the vendor that was chosen to fabricate the original contract's six frames was the Budd Company in Philadelphia. They also had bid on and won the contract to fabricate the initial six pairs, or twelve optical bar tubes (only the tubes, no other internal parts). The optical bar tubes had such large significant areas they also had to be buffed and polished to a low emissivity.

It was necessary to separate the two contracts (the frame and the tubes) because it was possible for a clever person to put together the fact that the two tubes could be mounted somehow to the frame. Someone could also figure out the included angle between the tubes and perhaps assume that this was going to be a stereo camera. Therefore in order to try to keep the two separate, the contract for the two tubes was written for The Boller and Chivens Company in Pasadena, California. The Boller and Chives Company was a telescope manufacturing company that happened to be owned by Perkin-Elmer.

When I visited the Budd Company to discuss technical problems and progress on the tubes, I had to appear to arrive in Philadelphia from California. Another method of evasion was that all of the correspondence relating to the tube was sent from Hank Brockschmidt, a Boller and Chivens Vice President in Pasadena to Budd and all of the mail from Budd was sent to California, and was then forwarded by Hank Brockschmidt's office to me at Perkin-Elmer in Connecticut.

All of this subterfuge was done for security reasons. Two of Budd's senior executives were cleared to the high security level required to know about the program. They succeeded in keeping the two programs separate. Interestingly, Brockschmidt was not cleared and never knew the contents of the mail. He just followed direction from our corporate management to forward the mail. Many years after the Hexagon program ended, I met him at a company get-together. I told him the story and we had a good laugh over it.

After the first six launches the fabrication of subsequent frames and tubes was awarded to different vendors. The frame was awarded to the Fairchild Corporation on Long Island, New York. Frequent trips had to be made there in order to review manufacturing plans or problems. Neil being the project engineer on the frame got to do that travel. He was lucky because Fairchild, being an aircraft company, had their own helicopter. Our company was about half a mile from the local airport. So when a trip to Fairchild was required, Neil would call his contact at Fairchild and a helicopter would be sent out to get him, and eventually bring him back the same day. The one-way trip took slightly over one hour and so Neil had the pleasure of an easy trip. He was able to enjoy sightseeing the scenery below as they flew over Connecticut, then Long Island Sound.

THE TAKE-UPS

Unlike reconnaissance cameras of today that transmit their photographic data to Earth digitally and in real time, the Hexagon system depended upon retrieval of the film. The technique for this had been used for many years. The main difference with this camera was that the weight of each film capsule was much more than ever, as there were thousands of feet of film in each.

After the photographic process occurred in the TCA, the film was transported into the forward section where it was placed into "buckets" or reentry vehicles that resided in tandem in the nose cone of the vehicle as shown in Fig. 12.2. There were four such buckets, and each was ejected from the vehicle and returned to earth with its "payload" of exposed film. Take-up assembly (TUA) number 1 was the most forward in the nose cone and was ejected first. All recovered TUAs were refurbished, cleaned, and reused.

The TUA that was spooling film rotated when the camera was operating and wound the film onto a core. When the film stopped moving in the camera,

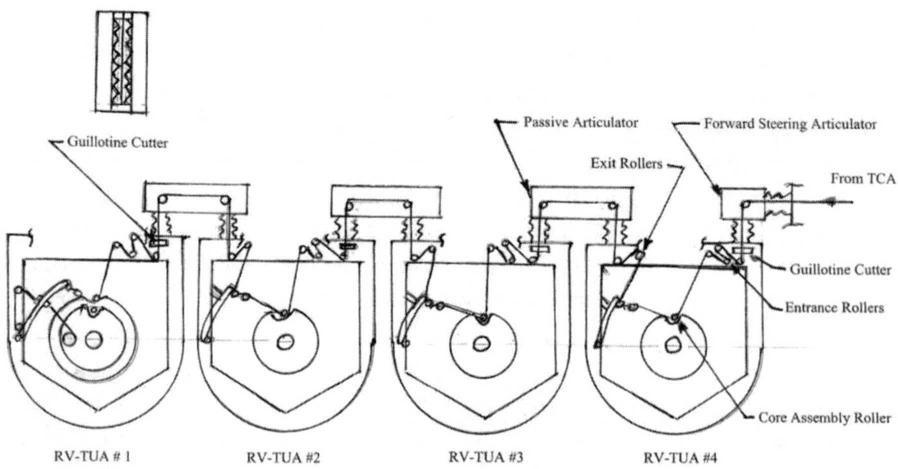

Fig. 12.2 Forward section of the take-ups.

the TUA was also commanded to stop. The TUA then waited for the next film turn-on operation.

After TUA 1 was full and with the core brake engaged, the TUA 2 was powered and its core rotated drawing film from the midsection and spooled several wraps around the core. The film entering TUA 1 was then cut by a guillotine-type cutter in the reentry vehicle and TUA 1 would be ready for the return to Earth. Transfer from TUA 2 to 3 and so on was accomplished in like sequence. The guillotine cutter is shown schematically in Fig. 12.2.

Each TUA consisted of a lightweight structural assembly that had two core assemblies sandwiched between three aluminum spoked frames, with a hollow central hub, all mounted axially on a stainless steel hollow shaft. Each TUA, shown in Figs. 12.3 and 12.4, contained the following major components:

- Roller bracket assembly
- Core assembly
- Builder roller (BR) assembly
- Brake assembly
- Deorbit pin assembly

Don Fish was the original design engineer on the take-up assemblies. He was succeeded by Ted Urban (Fig. 12.5), a member of the Quality Assurance department who was intimately involved with the workings of the take-ups. Ted eventually went to California and became part of Perkin-Elmer's West Coast Field Office quality engineering staff.

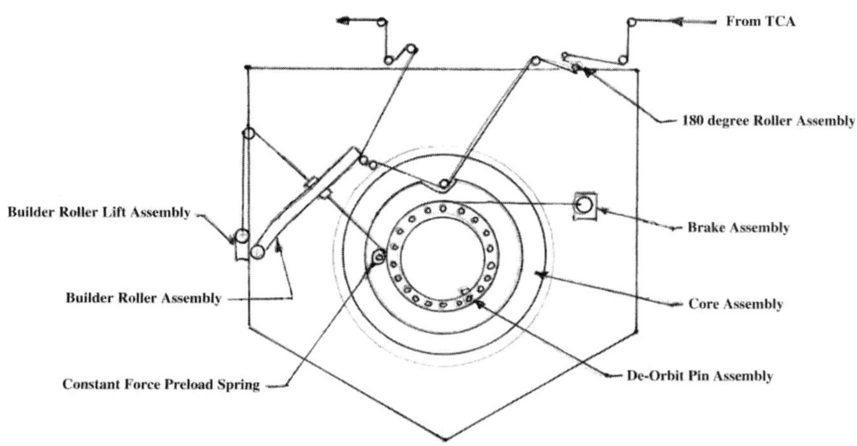

Fig. 12.3 Schematic of the take-up assembly.

Fig. 12.4 Take-up.

ROLLER BRACKET ASSEMBLY

The input and output roller bracket assemblies contained several 0.5-inch rollers and were mounted to the outboard and inboard structural frames. They guided the film from the TUA entrance to the core assembly roller and then to the TUA exit. All rollers were aligned parallel to each other and to the core within one thousandth of an inch.

CORE ASSEMBLY

Each TUA had two cores, called "A" and "B" corresponding to the A and B cameras. The cores were motor-driven and wound film for each camera. They turned in opposite directions to balance rotational inertial torque loads and provided momentum balance.

Each of the two rotating core assemblies contained a 16-inch diameter hollow beryllium ribbed drum. Angular contact bearings were positioned at each end of the drum and interfaced with the stationary 14-inch diameter shaft. The rotating parts of the DC brushless motor were installed in the hollow core. Their stators interfaced with the fixed shaft. A 0.5-inch diameter roller assembly was installed in a recess in the outer surface of the cylinder so that the roller diameter was flush with the cylinder surface. The motor and encoder electronics were installed in the hollow shaft.

BUILDER ROLLER ASSEMBLY

As the film was being wound onto each drum during photographic runs, it was "ironed" down with a builder roller that forced out air trapped between

Fig. 12.5 Ted Urban, take-up design engineer.

layers of film on each drum. This ensured that the many layers of film were pressed down uniformly to create a film stack having uniform density and that the thousands of feet of film on each drum lined up in a uniform cylindrical shape and maintained its internal tension thus preventing "loose" film and telescoping of the film under axial loads during reentry.

The builder roller consisted of an aluminum arm, a pivot assembly at one end of the arm and a set of 180 degree rollers at the other end where the last roller contacted the outer wrap of film. The angular position of the arm of the builder roller was used to determine the diameter of the outer wrap of film with a potentiometer attached to the rotating pivot shaft of the builder roller.

When the project engineer Don Fish retired, I organized a farewell party for him. As a parting gift I found a spare, slightly damaged builder roller arm and attached one end to a plank of wood and at the other end I attached a roll of toilet paper. I put hooks into the wood so that it would be hung on a wall in the bathroom. All those in attendance knew that Don had been responsible for the design of the whole TUA and they laughed loudly when they saw this gift. I never found out what his wife thought about it or what he did with it.

During tests of a TUA it was discovered that when there were many layers of film wound on the core the stack became slightly tapered (meaning the outer diameter of the film stack was higher on one side of the film than on the other). In time, this taper across the outer film layer of the stack became 0.025 inches or greater. When this happened, the photo operation would "crash," meaning that the film spewed to the side of the stack and caused a jam. We identified this problem to Kodak who took steps to fix this.

The film roll was comprised of many separate film lengths that were sequentially joined together. Kodak implemented a method that measured the thickness taper across the width of each piece of film. The nominal thickness of the film was 0.002 inches. If one edge was 0.002 thick and the other edge was 0.0021 inch thick, this created a taper of 0.0001 inch. If 250 layers of this were wound (a full TUA had many more layers of film), the taper across the film width for the whole stack would be 0.025 inch, thus causing a problem. As a solution Kodak measured the taper in each of the film pieces and butted positive and negative tapered lengths together, thus reducing the average taper to an insignificant amount. This was also found to improve film flatness in the focal plane.

Two other steps were developed by us to prevent the stack from crashing during operation. Most of the design refinements relating to film were developed during tests in the engineering model. Sam Palasciano, Don Cowles, and Ed Matthews were the three engineers who performed the tests and came up with relevant solutions. These were to install two additional rollers and a pivot for these rollers in the film path just prior to where the film entered the core, as shown in Fig. 12.3. The additional rollers caused the film to be wrapped 180 degrees around the rollers instead of just 90 degrees around one roller, thus providing more stability. The pivot acted as a steerer in order to feed the film onto other rollers uniformly.

The builder roller was locked in the fully raised position during launch and held by a caging pin. After launch a solenoid was energized, pulling the caging pin back to disengage the builder roller that allowed it to be lowered to contact the film's outer wrap.

During photographic runs, the builder roller applied a steady force to the outer wrap. A wire attached to a spiral negator spring (constant force) located on the support frame pulled the arm down causing the force. During inoperative periods the builder roller was raised approximately 1/8 inch above the outer film wrap.

BRAKE ASSEMBLY

When not in operation, the core assembly was held stationary by a spring-loaded stainless steel brake band attached to the outboard structural frame. It engaged a polyurethane surface on the core drum over an angle of 270 degrees thus providing the braking action.

Deorbit Pin

The function of the deorbit pin was to lock the core assembly to the TUA when spooling was complete. This prevented rotation and possible unspooling of the film when subjected to the reentry loads and the shock of being caught by the parachute snatch. There were a series of 0.25-inch holes around the periphery of a ring on the drum. A pin/solenoid assembly was mounted to the support frame. Upon energizing the solenoid, the spring-loaded pin was released to engage one of the holes. A slight rotation of the core was usually required to engage a hole.

After its return to Earth, the TUA was transported to a Kodak facility where despooling and analysis of the photographs took place. The empty TUA was then returned to the factory at Perkin-Elmer, tested, disassembled for inspection, then reassembled and reintroduced into the next system for further use.

In the next chapter, Hexagon's reliability is discussed.

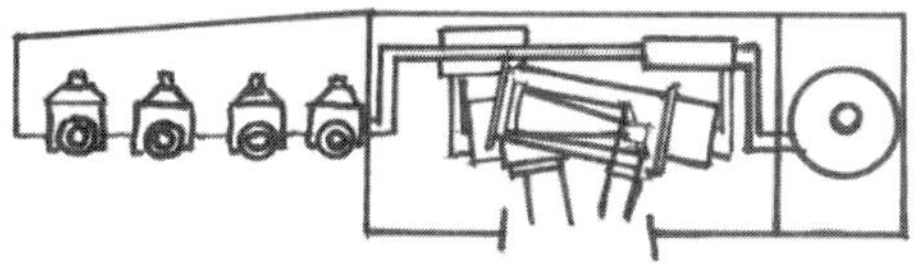

Chapter 13

Reliability Aspects of Hexagon

Oscar Berendsohn

*Introduction**

Reliability as related to Hexagon was important in several ways. It involved mathematics and statistics in predicting probability of failure of electronic components based on their life test results and published failure rates. It also involved materials such as metals, plastics, lubricants, and cements and their strength, stiffness, and thermal properties in order to address mechanical failures in various structures.

The Reliability department was made up of experts in these fields as well as laboratory technicians to perform all kinds of tests. My friend and colleague Oscar Berendsohn discusses some of the problems and his experiences on the program (Fig. 13.1). He was born in 1924 in Hamburg, Germany. He was an old codger; but only in terms of years, not body and spirit. Just like me he was a survivor—he barely made it out of Germany after the 1938 Kristallnacht (Crystal Night) terror with his family and spent the next 2 years in Honduras before being able to immigrate to the United States.

After World War II he had a number of interesting jobs for a few years, none demanding much enterprise or skill. Among them was a stint in a malleable iron foundry where he learned how to sweat for his wages in the true meaning of the word. It was probably the foundry more than anything else that finally persuaded him a better future was to be had with a better education; and so, after having served in the army during the Korean War he enrolled in the Polytechnic Institute of Brooklyn where he graduated with a bachelor degree in metallurgical engineering. It was the turning point in his life.

It was the age of Sputnik, the beginning of the space age, and he found himself as one of its first devoted followers. It seems that he had a talent for discovering the causes of electrical and structural malfunctions. These are of major concern in any space effort because no repairs were possible in orbit.

Copyright © 2013 by the American Institute of Aeronautics and Astronautics, Inc., All rights reserved.
*Introduction by Phil Pressel.

Fig. 13.1 Oscar Berendsohn, Reliability department.

By the time he was interviewed in June 1967 for a job at Perkin-Elmer, he had already accumulated about as much experience in the area of malfunction analysis as anybody in the field since it was relatively new at that time.

OSCAR BERENDSOHN'S STORY

BEGINNINGS

The person who interviewed me for my job was Robert E. Smith, known among ourselves as "Smitty." He was a man of about 40 years. With slightly graying dark hair and of medium build, he looked at me during the interview from behind thick horn-rimmed glasses with never a smile on his face.

Once you got to know him you found him to be a friendly person and a talented professional. Like me, he had studied metallurgy. At that time he was a section manager in the Reliability department. He seemed satisfied with my credentials. After a lengthy interview he indicated that I was the front-runner for the job. "I must tell you though," he concluded, "your ability to work on this project depends on obtaining the necessary security clearance." I thought that should be easy and I told him so. "This is different in many ways," he responded. "There will be an extraordinary inquiry into your past. Is there anything you might want to tell me now?"

"No, my conscience is clear," I replied. "I have not robbed anybody for quite a while," and smiled at him, but this did not seem to amuse him. He said only that I would hear from the company in a few days, and that was the end of my interview.

I waited anxiously for the reply. In truth I knew, but had not told him, that something of the importance of the project had somehow slipped out. At my last job, my boss had received an urgent phone call in my presence from his stockbroker telling him that Perkin-Elmer had just landed a large government program but nothing of its nature was known.

A few days after the interview I received a telegram that offered me a job as an engineer in the Reliability department of the Systems Operations Division. I was "in" as the saying goes, and I was determined to stay in. In my heart I knew that this was the opportunity of my lifetime.

As everyone else who was waiting for their security clearance, I was assigned with hundreds of people to the tank. It was a large open area that at one time had been a clean room. The first few days dragged by in monotonous routine filling out the necessary forms for obtaining my clearance. I did the best I could, but after so many years, I could not be sure if everything that I put down concerning my past was correct. Were any relatives in East Germany? No, I entered and that was true, but of course it could have been different through no fault of mine.

Achieving Low Failure Rate

We also received training in theoretical reliability engineering. This branch of engineering is based on probability theory. While mechanical systems lend themselves in general to a very high degree of predictability this is not at all true of electrical systems unless extraordinary precautions are taken to determine and control the failure rates of each type of component whether it be a relay, a microcircuit, or a printed circuit board. The failure rate in service of each part must be known in order to calculate with complex equations of what must be accomplished in order to meet the desired operational degree of success. In our case it was our goal it have a 95-percent rate of success; or, to say it another way, to achieve full or any success, the probability that only one in twenty systems could fail.

In order to obtain this low failure rate, all parts had to be "burned in," meaning that we put them on a rack, activated them, and let them run for a number of hours thus eliminating all the weak ones among them, known as *infant mortality*. In some cases even this was not enough, and we had to set up special production lines at the vendors to make ultra-high quality parts. This was particularly true of electrical relays.

If the required failure rates were still too high, redundancy had to be designed into the circuit so that it functioned even if a failure occurred. In addition, a failure mode electrical criticality analysis (FMECA) was performed on the entire system to determine the severity of any particular mishap. If all of this sounds very expensive, it certainly was, but was unavoidable for the type of system we built, and our attention to reliability paid off in a big way.

I believe the Soviets failed to appreciate the significant difference between quality control (inspecting parts) and reliability (calculating required performance). They did not apply sufficient mathematical analysis to their space effort reliability requirements, and as a result experienced dramatic failures. We will see later on in this chapter how unacceptable electrical component failure rates could have ruined our system.

Receiving Security Clearance

During the next few months I began to receive curious glances from some of my neighbors in East Meadow, Long Island where I lived at the time. It seemed that mysterious strangers had come to their doors asking about a certain Oscar Berendsohn and what did they know about him. Was he ever drunk? Did he belong to any unusual organizations? Who were his friends, and did he have any large debts? Certainly, I must have committed some dastardly crime for the authorities to be so concerned about me.

Whatever their answer was, it must have satisfied the folks behind the curtains because in a relatively short time, about four months, I received my clearance. What a day for myself! I had all sorts of ideas as to the nature of the program. I was told to see Robert Markin, the head of our Security department at that time. There I was first asked to sign a document to the effect that I would not disclose any information to any unauthorized persons. In addition, I must notify Security of any trip abroad. Any travel behind the Iron Curtain was strictly forbidden as well as any trip to any other Communist nation. West Germany was all right though.

Most important, the customer wished to remain anonymous and his identity would never be revealed to anyone under any circumstances. He would be referred to only as "the customer," period! If anybody asked me what we were working on, I should tell him or her it was "proprietary."

I signed the required forms and then was given a thick volume of information concerning details of the Hexagon program. For the first time I learned its nature. A reconnaissance satellite, but the devil is in the details as they say. This one was simply mind-boggling in its complexity of construction and intended mission. Much of the information was outside my field of expertise, but some of the mechanical details were obvious—and staggering—in scope. The film was to be transported at breakneck speed supported by a thin layer of nitrogen between it and some of the rollers on its way to the slit-and-shutter assembly in the focal plane area. I learned later on that the method of taking photography was made possible by the brilliant concept of our mechanical engineer Don Cowles. Would it work?

The nitrogen was stored in two large spherical tanks pressurized to such a degree that an explosion would have leveled our building and blown us, the ministering angels, to kingdom come. Since I value my earthly existence at least as much as angels, I knew then and there that their reliability would be one of my primary concerns on the program.

There was simply not enough time to retain most of the information I was given to read. Overwhelmed by it all, I handed the file back after about an hour or so to a Security fellow and it promptly disappeared into a large safe. "Follow me," said Bob Markin, and he led me to a part of the building I had never been to before. He paused before a cipher-locked door and punched in the combination. We were about to enter the inner sanctum of the Hexagon program.

Inside I stared in awe at a full-size model of Hexagon. Markin explained some of the details to me about the two rotating optical bars that would give the system its stereo vision, the two massive drums with film, the four take-ups, and the mechanism for cutting the film once a take-up was filled as well as the slit-and-shutter mechanism. For the first time I had a look at the pressure vessels. In order to save weight their walls were quite thin, requiring a construction material of extraordinary strength.

After my tour of the Hexagon was complete my head spinning. But it was not from the enormity of the program alone. In all seriousness I could not see how we could ever make something this complex function correctly at all—let alone make it perform as specified. "The folks who came up with this are having pipe dreams" went through my head. The system was simply too complex, too many parts, too many "ifs." One broken spring, one faulty switch, one seized bearing, and it would all be over. The whole glorious thing would be nothing but so much junk floating in space. In the weeks to come I exchanged my views with many of my coworkers. We all agreed. It will never work! Happily we could not have been more wrong.

MATERIALS SCIENCE SUPPORT

My primary function was not failure investigation, but materials science support. We used a lot of exotic materials. The system used all sorts of different stainless steels: titanium, lots and lots of beryllium, aluminum alloys of many types, brasses, Invar, magnesium, and beryllium copper. The engineers responsible for the major assemblies made the decision for most of these materials because they knew the specific requirements better than I did. I would then review the drawing for any mistakes in specifications. Welding or brazing, finishes such as platings, solid lubrication, and heat treat requirements were a different matter for which I would usually end up making recommendations or actually make the selection, and in some cases required that only vendors that had gone through a qualification process were to be used.

As important as this function was, it was fairly routine as far as I was concerned. It was when test failures occurred that I had to earn my keep. In part this was because I had to determine at once if test conditions had been unrealistic, if someone had accidentally broken the part, if a quality defect existed, if a basic error in materials selection was responsible, or if the design might be the culprit.

Some of the more interesting types of problems that the Reliability department staff investigated fell into the following groups:

- Structural failures
- Dimensional stability problems
- Electronic failures
- Contamination problems
- Coating and finishes problems
- Joining methods (such as welding, brazing, and soldering)

In the beginning, the main drive shaft for the take-ups was made of beryllium, a lightweight material with exceptional strength and stiffness, but also extremely brittle and ultra toxic if its dust or vapors generated during machining are inhaled. Three times in a row a shaft had broken during testing at RCA near Princeton, New Jersey. RCA was a subcontractor to us at Perkin-Elmer on the take-ups. Later we retrofitted the take-ups ourselves.

The last failure occurred just before quitting time one day. My department head, Stanley Karachuk caught me, as I was getting ready to leave. "Not so fast, Oscar," he called after me. "You have to go right now to Hazleton, Pennsylvania. There will a meeting tomorrow morning. Another take-up drive shaft has failed. The meeting will be at the Beryllium Corporation. Be there. It will be a classified meeting in some respects so before you go see Security."

I called my wife and told her I would be gone for 2 days. By now she was used to some of these unusual trips I made from time to time. Security informed me that I was to attend the meeting incognito. My nom-de-guerre would be Oscar Braun, German accent and all. The next day we held the meeting and the customer and RCA were there. I was introduced as Oscar Braun. "Are you related to Werner von Braun?" asked one of the fellows. "Yes," I replied with a straight face. "He is my uncle but I dropped the 'von' from my name. I keep an autographed picture of him on my desk." I received a look of deep admiration.

The fractures were attributed to a stress concentration, a design problem in the form of a keyway. However, subsequently we decided to change the material to aluminum alloy 7075, a high-strength material and not brittle at all. But of course there was a weight penalty as beryllium is 30 percent lighter than aluminum.

Spherical balls contained high-pressure nitrogen. From the very beginning we knew that the pressure vessels that held our supply of nitrogen had to be treated with the greatest of respect because not only the mission, but also life and limb were at risk. They were pressurized to 3000 pounds per square inch (psi). Since one atmosphere is 14.7 psi, these vessels had to withstand a little over 200 atmospheres, a horrendous pressure. The reason for this high pressure was to make sure that there was enough nitrogen for the whole mission.

Originally they were fabricated by *cryoforming*, that is, the two hemispheres making up the shell were shaped at −100° F. This greatly increased their strength and they worked well without any problems. After the first group had been used up, we placed an order for a second lot and that is where the trouble began. The vendor could no longer make the vessels to the required strength level because some key personnel had left the company and taken the required expertise with them.

We had to choose a new material. I had advocated titanium as a replacement, but the head of the Engineering department, Charlie Karatzas opted for an ultra-high-strength steel called Maraging 200. This material had many good properties besides strength. It was quite tough and, most important, was readily weldable. The order was placed with the Walther Kidde Company in New Jersey. All went well at first. Because the finished sphere could not be x-rayed to look for hidden defects (it was impossible to place the x-ray film inside the sphere), we had to resort to ultrasonic testing, but there was a problem. The resolution of this method is about 3/32 inch, meaning smaller defects could not be detected.

Fortunately for us, the science of fracture mechanics had been developed in the meantime. One of its advantages is that it can predict the allowable flaw size for a chosen part from a property called *fracture toughness*, which is fixed for a given metal or alloy. In turn, this means that it is possible to examine the part to make sure that the allowable defect size has not been exceeded. Unfortunately, in our case, the allowable defect size for Maraging 200 was close to the observable size of 3/32 inch. As a consequence, every vessel was proof tested by pressurizing it with water to 10 percent above the required 3000 psi. Everything seemed fine, but not for long.

We discovered that this type of steel is prone to a failure mechanism called *stress corrosion cracking*, one of the most dangerous types of failures because it can occur randomly and without warning or observable crack development. It is caused by a combination of high stress and moisture (provided by the proof test conditions). Cracks occur at the microstructural level at selected points only and produce enormous local stresses unobservable by any means until failure occurs. Proof testing does not help here as we would never know if a tiny crack was growing and the part would fail later on spontaneously. But something had to be done.

There was a new method called *acoustic emission testing*, which is based on the sound detection of the tiny noises given off when a miniscule crack grows internally in a metal assembly while under stress. Its use is generally restricted to atomic reactors as well as chemical and other pressure vessels due to the high cost of the technique, about $5000 for each test. I approached my bosses, who supported me, and then the Program Office to carry out the test on all vessels. Initially, I met resistance there to the idea, but finally received permission to test a spare vessel to failure (to determine the safety margin) using this technique.

Wyeth Company near Redlands, California carried out the test for us. The test was successful. The pressure vessel did not exhibit any evidence of stress corrosion. I had hoped that the Program Office would reconsider their opposition to testing all the remaining pressure vessels, but I was wrong. They were not. I had a bad taste in my mouth years later when the last mission exploded shortly after launch at Vandenberg Air Force Base. Subsequent investigations place the blame elsewhere.

AVOIDING DIMENSIONAL CHANGE

In an optical system of the accuracy that was required of us, dimensions take on a new meaning. Optical component dimensional requirements for the Hexagon system were given not in inches, but in fractions of a wavelength of light or in angstrom. The wavelength of light at a particular frequency is 26 millionths of an inch. It was therefore important that the metal parts that held the optics of other precision components not change their dimensions over a long period of time.

Such dimensional changes occur in many ordinary and also exotic materials, but are usually of no concern because they are very small with the exception of shrinkage stresses in welds. In our case, more often than not, we had to pay attention to the potential of time-dependent dimensional changes that could cause havoc with our optical and other precision requirements. As a consequence, we had to develop heat treat and machining sequences that would eliminate internal stresses or time-dependent metallurgical changes.

The first of these problems surfaced early on. A large support structure (a steel weldment) was fabricated for us by Tower Iron Works in Rhode Island.

The rush was such that we had to get there in the middle of the night to instruct the vendor on his procedures. The assembly was much too large for annealing (heating to relieve internal stresses and then slowly cooling the part).

There was only one alternative left to us: perform vibrational stress relief. A vibrator oscillating at a critical frequency was attached to various parts of the assembly and run for up to 1 hour. Experience has shown that this unorthodox method, while not perfect, removed most of the stresses to a degree that made the part stable for our application.

The most difficult and expensive procedures involved in the Hexagon system were for the rollers supporting the film closest to the focal plane structure. All rollers were made of beryllium. These particular ones had some dimensional tolerances of 50 microinches and had to maintain this tolerance during their service life.

We met the requirement by machining the parts in sequenced steps, chemically etching the parts to remove machining stresses and micro cracks induced by machining this ultra-brittle metal. Once the part was within 2 or 3 mils (one mil = 0.001 inch) we would then stress-relieve it. The total required sequences of machining, etching, and stress-relief were usually more than six separate steps. As a last step, we would temperature-cycle parts between the expected temperature extremes to avoid dimensional changes that might occur during operations over a long period of time. Similar procedures had to be followed for essentially all beryllium parts that we used. Obviously this makes such parts incredibly expensive. Add to this the price of beryllium that was about $1000 per pound in those days, as compared to $1.50 per pound for aluminum. However, without the use of beryllium we could not have squeezed the required components into the volume allowed for them not to mention the enormous savings in weight.

We also made extensive use of a steel alloy called Invar. This alloy made of iron and 36-precent nickel is unique in that it expands or contracts very little when heated and cooled within the range of our system's temperature extremes. Because fused silica and other glass materials used in the camera had similar expansion/contraction properties, Invar would be ideal to support the optical components. Invar however was a fickle material whose behavior on heating or cooling is influenced strongly by traces of impurities, mainly carbon, as well as stresses from machining and Curie temperature effects (temperature at which Invar becomes nonmagnetic) during annealing to remove machining-induced stresses. The complexity of Invar was such that we might have had serious problems had it not been for a doctoral dissertation on the alloy by an MIT professor. His data allowed us to specify the proper heat treatment procedure for those parts.

LIGHTS OUT: AN ELECTRONIC PROBLEM

One electronic problem that stood out from all the others was the miniature lightbulbs that were a vital part of our encoders. The encoders I refer to have nothing to do with cryptography. They are electromechanical devices that control the movement of an electric motor. A glass disk with thousands of tiny metallic lines moves in front of a light beam. These lines interrupt the beams from the light sources to generate the ones and zeros that are the digital signals that provide the following functions: motor commutation, shaft angular position, shaft angular velocity, and trigonometric functions of shaft angle.

The light beam was provided by a miniature lightbulb not similar but exactly like those used for some toys. Therein lay our dilemma. A single failure of one bulb could have crippled our mission. The system used more than forty of them, one to each encoder. It was a nightmare for those of us familiar with the problem. The reason was simple: No matter what we tried, we could not persuade any of the four or five manufacturers to make us special high-quality bulbs. "Buy them as they are or leave us alone" was the answer by all of them. We offered to pay $50 per bulb instead of the pennies they normally cost if they were made to our specifications. We were laughed at or simply rebuffed. This included General Electric. I tried to persuade the head of their lighting center in Nilo Park. He told me I was wasting his time for the few bulbs that we needed.

The upshot of all this was that we were forced to buy literally bushels of bulbs, triple inspect them for quality, and then burn them in. Out of a lot of 2000, perhaps 30 or 40 might pass our screening. Tiny glass cracks, distorted filaments, faulty hermetic seals, and poor filament attachments were the main rejection criteria.

The filaments consisted of a tiny tungsten wire about one mil in diameter, which is one third the thickness of a human hair, and all that had to happen to ruin our mission was for one of these tiny whiskers to burn out. End of performance, the curtain would come down on us. Not only that but a distortion of the white-hot filament of only 15 degrees was just as bad as it would cause the light beam to miss the slot through which it had to pass.

Boller & Chivens, one of our divisions based in South Pasadena, California built the encoders for us. Initially Dynamic Research Corporation in Massachusetts made the encoders. David Burch was in charge of product assurance at Perkin-Elmer and over the next few years he and I became good friends. Both of us shared the urgency of the situation, and tried all sorts of maneuvers to find a better solution to the lightbulb problem, but to no avail. We were stuck with the same toy bulbs. Sooner or later we knew our luck would run out as it did for two Landsat satellites that relied on these bulbs.

When we saw the news of those failures Dave and I looked at each other and had the same thought. Fortunately light-emitting diodes (LEDs) improved with respect to light intensity a few years later so that we were able to substitute these for the toy bulbs, and Dave and I could breathe a sigh of relief.

CONTAMINATION

The Reliability department was also tasked to confront contamination problems and these fell into different types. The first and most urgent was the need to eliminate particles of all types and sizes down to the tiniest from our optics and for good reasons. Wherever there was a particle there would be a defect in the photograph.

When the mirrors were cleaned after assembly, the rinse solution was retained and then drawn through a micropore filter. Any residue on the filter was analyzed to find the source of the particles.

The same was done for all assemblies when they were vibrated to simulate launch conditions. In the beginning, before the source of these contaminants was known, the debris removed after vibration was sometimes astonishing. Found were human hairs, little bits of wires, washers, even the occasional moth, but the most serious of all of them were frequent Heli-Coil tabs. A Heli-Coil, an invention of a company of the same name, is a spiral wire shaped to a specific diameter and number of turns per inch to replace threads in materials such as aluminum when there is risk that the bolt could seize or pull out of the threads. The Heli-Coils were inserted into a predrilled hole with a special tool. Once fully inserted, the end of the Heli-Coil consisting of a tab was broken off.

Once the source of the contaminant was known, it became a matter of holding meetings with manufacturing and quality control personnel to institute measures to eliminate these sources. The success was astonishing. In a short while the moths, tabs, washers, and human hairs all decided they would go elsewhere and not visit us anymore. The optics and assembly were essentially free of contaminants.

Evaporation of volatiles was probably one of our greatest concerns because the Hexagon in service would be exposed to a hard vacuum; that is, a vacuum below 10E-6 torr (torr equals 1 mm mercury or 1/760th of an atmosphere) where any volatiles could condense onto the cooler optics and cause obscuration.

A new and unusual source of contaminant came as a complete surprise. One day acetic acid was discovered on the optics. Acetic acid is the ingredient that gives vinegar its peculiar smell. In our case it was not only a source of corrosion, but more importantly in a vacuum it would volatilize and coat our optics with a layer of film. The source of the acetic acid was soon traced to a new type of silicone we were using. In this particular type (and in many others of similar composition), acetic acid is split off as part of the polymerization reaction; that is, when it forms chains of molecules. The polymer chemist refers to this as a *condensation reaction*. We replaced the silicone with a type that splits off alcohol. The alcohol, of course, promptly evaporated.

We also came across contaminant that we had to analyze just to figure out what it was. I have mentioned the system of supporting the film on jets of nitrogen while passing through the focal plane assembly. These jets of gas came out of aluminum assemblies shaped in cross section like a "D" and for that reason were referred to as D bars. The nitrogen jetted out of many tiny holes in the rounded portion of the D. The opening for the jets was only about 0.003 inches in diameter, the thickness of a human hair.

We had no sooner started production and testing of these bars when trouble reared its ugly head. Many of the holes were plugged up. The faulty parts were sent to me for evaluation. When I opened them up and flushed the interior for particles I discovered a number of small white specks. These were analyzed by what was then a relatively new method, namely scanning electron microscopy (SEM) with a microprobe attachment. The microprobe sends out a beam of electrons that bombard a microscopic area of the sample. The elements present generate x-rays as a result. Recording the energy of the x-rays (or the wavelength) identifies the elements present and their relative amounts. In this way I discovered that the white particles consisted of brazing flux used in assembling the D bars. The answer to our problem was simple. I spoke to Don Cowles, the inventor of the twister, regarding our findings and we agreed to change the assembly method to welding that requires no flux. It worked to perfection. None of the D bars ever plugged up on us again.

It was also possible that a contaminate could corrode something. I had never been to the assembly area at Lockheed in Sunnyvale where the system was integrated and made ready for its trip. All this was about to change for a most bizarre reason. Somebody had accidentally released hydrochloric acid vapors into the air conditioning system at our assembly area. The fear was that the vapors might have attacked portions of the take-ups in the recovery vehicles. I was sent out to examine the four take-ups.

I must say that I was quite excited to see the entire assembly for the first time. The sight of the huge craft sitting there in the gigantic assembly area was simply awesome. Examining the take-ups in the recovery vehicles was a challenge even for a contortion artist, but in the end I managed. Our people had worried for nothing. There was not a trace of corrosion to be found if my life depended on it. As a precaution I requested that certain parts made of beryllium copper and a spring made of music wire (a high-strength steel wire) be replaced. So it was done. The parts were sent to me for further study. We had worried in vain. But better an ounce of prevention that a ton of debris!

PLASTICS AND LUBRICANTS

Metals with rare exceptions, such as mercury or when in their molten state, are not subject to outgassing or evaporations under conditions just described. But plastics and lubricants were a different matter. For many plastics such data had already been established from other space programs, but in many cases we had to obtain our own data by measuring the weight loss of these materials before and after exposure to a vacuum.

It was usually relatively easy to find a plastic for our many applications such as printed circuitry and adhesives for bonding that was acceptable. But for lubricants, the choice was extremely limited and this was significant because the Hexagon system contained many moving parts including hundreds of bearings that needed lubrication.

There were very few—and very expensive—oils or greases such as Braycote and Versilube that were suitable for us. In many situations however we resorted to what is known as "solid" lubricants. These do not contain grease in the normal sense, but are films containing a plastic binder to hold them in place. Virtually all of them consist of a lubricant called molybdenum disulfide (MoS2) plus some kind of binder. Teflon may be used for applications involving impregnation of anodized aluminum wear surfaces.

Most existing solid lubricants were proprietary by the companies that coated our parts. One of these was a lubricant called "Lifelube." It was used of all places on the seal for the optical bar. "Deathlube" would have been a more appropriate name. It worked well for a while but then serious problems developed. The wear rate became excessive. Something was obviously wrong. We held discussions with the supplier. A dignified-looking gentleman sporting a white mustache by the name of Dr. Hopewell represented Lifelube at the meetings. When questioned he insisted that nothing had been changed in the process. Eventually we discovered that a critical step had been changed or omitted—surface preparation. A very fine sandblast or vapor honing was required. The tiny holes this forms act as pockets to retain the lubricant. Once this method was reinstated the coatings worked well. Before too long though, Lifelube went out of business and we had to change to another coating for our parts; we used Microseal 200 supplied by E/M Lubricants.

CONCLUDING REMARKS

"There will be a meeting at 1:00 p.m. in the cafeteria," the loudspeaker blared suddenly one day in October 1978. "Everybody will attend." We were puzzled. What could possibly have happened that was this urgent? Maybe the contract was canceled. All sorts of wild rumors spread with the speed of greased lightning and as usual everybody guessed wrong. Our blood pressure soared. At a few minutes before 1:00 p.m. a mad dash for the cafeteria began. For most of us it was standing room only.

At shortly after 1:00 Paul Petty, our general manager, mounted the podium. He checked with Security to make sure no uncleared persons were present. "We have been informed by Security that President Carter will give a speech tomorrow," he began. "In his speech he is going to talk about the surveillance satellites of this country," he continued. "I must make it clear to everybody that this does not mean that you may disclose anything whatsoever about our program. As far as our security requirements are concerned nothing has changed. Any questions?" There was a sullen silence. I believe that for the most part we resented such a disclosure. The U.S. president admitting this fact could gain nothing. I don't know what President Carter had in mind but as far as I was concerned (and most of us I believe agree), such statements were foolish and ill conceived. A cloak of silence must of necessity envelop intelligence as far as official statements are concerned.

Eventually the program like any other came to its end. It went out literally with a bang, but not of the type we had hoped for. The last launch in our then current contract exploded a few hundred feet above the launch pad at Vandenberg Air Force Base. Nonetheless as a whole the program was, to use an old phrase, a smashing success.

I don't believe there is a single soul among us who worked on this program who did not take extreme pride and satisfaction from the success of the 19 missions that were dramatic in their accomplishments. On my last day before retiring from the company, I visited our situation room. There on the far wall was superimposed a giant map of the world where we could follow the present position of our system. "Well done," went through my mind. We have helped make the world a safer place for us, for our children and for the generations to follow.

The next chapter discusses manufacturing and testing.

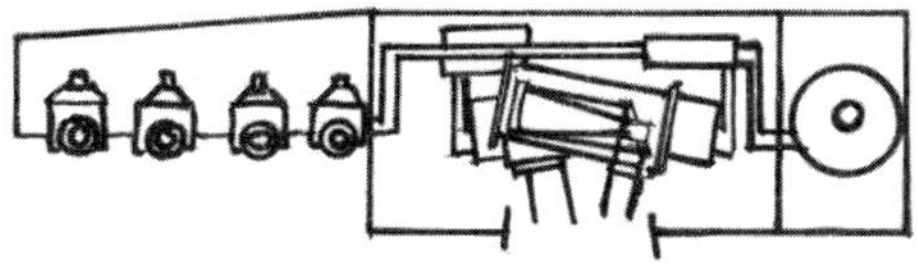

Chapter 14

MANUFACTURING AND TESTING

PHIL PRESSEL

Every single piece of the system, from lenses to housings, from cables to nuts and bolts, had to be fabricated or purchased and then put together. Many assemblies had to be tested. Thousands of parts made up this camera system. Most of the hardware was obtained from outside the plant. We did the design. Others fabricated the parts and shipped them to us; for example, the glass for the optics was purchased and then cut and polished in-house. An exception was for some of the electronic boxes where outside companies designed and fabricated the circuitry to our specifications. They had the manpower and manufacturing expertise. All parts went through a meticulous quality inspection program. Electronics had to be tested prior to shipment to us.

Most of the mechanical assemblies consisted of hardware that was very sensitive to contamination of any kind, including dust. This included lenses, mirrors, delicate mechanisms, and many state-of-the-art ball bearings. These would all require assembly in clean rooms. The very large high-bay areas were *class-100,000 clean.* This meant that the area contained less than 100,000 particles sized 0.5 microns or larger per cubic foot of air. The smaller assembly areas were class 10,000 and within those areas there were many class-100 laminar flow workbenches. These were used to assemble the many mechanisms and subassemblies such as the platen, film drive, and looper. The complete mechanical and electrical assembly of these could take as long as several months each.

CLEAN AREAS

Prior to going into these areas it was required that everyone first dress in clean-room uniforms, booties, hats, etc. Then entry was done by going through an interlock chamber: once the entrance door was closed, the exit

Copyright © 2013 by the American Institute of Aeronautics and Astronautics, Inc., All rights reserved.

door would not open until a 1-minute air shower was completed. The air shower blew off all contamination particles from a person. There were two interlocked entrances. White suits were for those working in the class-10,000 area and blue suits were for those in the high-bay. Special sticky mats were stepped on prior to entering these chambers in order to pull off debris from the underpart of shoes or booties. Only fiber-free type of paper and pens were permitted in the clean rooms because they would not emit or outgas foreign substances or discard fibers. All incoming parts had to be cleaned with special materials or by ultrasonic cleaning machines, then bagged (some sealed in dry nitrogen) and tagged. If someone needed to touch hardware, they had to wear gloves.

In order to handle large and heavy assemblies or intricate mechanisms, special tools and equipment had to be designed. Some of these had additional impositions placed upon them as that they were going to be used in a darkroom. That was required when loading film and performing certain tests on the heavy and large supply assembly.

Highly skilled mechanical and electrical technicians were the ones who did all of the hands-on labor. Each subassembly had a designated manufacturing engineer who worked with the relevant design engineer in instructing or troubleshooting during the assembly and test process.

There were vacuum outlets in all the walls of the clean areas. Large hoses were attached to these and used to clean dust or debris generated during the assembly processes. When epoxy cements needed to be removed, a small 0.5-inch plastic tube was extended from the vacuum outlets directly to within an inch or two of the epoxy being cut or twisted off. The tube would immediately suck up the epoxy debris, thus preventing the debris from inadvertently entering the assembly.

*JACK WALSH'S STORY**

When I first was hired in 1968 they were looking for people who had project engineering management or manufacturing engineering experience. I was at Sikorsky and had a lot of planning background, so at Perkin-Elmer I ended up on the development model with George Henderson as program manager. It took probably 3 months to get cleared as by then they were pushing to get clearances through as fast as possible.

Meanwhile I started working in the uncleared area called the tank. We were given assignments, not knowing what the application of the hardware would be. I worked on some of the bearing modules (9 and 14 inch) that

*Jack Walsh was one of the Manufacturing department managers.

were fabricated at either the Speedring Company in Detroit or the Pioneer-Astro Company in Chicago. We shipped the bearings to them and they had to fit the bearings to the housings and do it in a clean-room facility. They had to match the housing diameters to the as-measured diameters on the bearings to make the appropriate fits. Then they had to assemble them and perform torque trace measurements. That whole technique all had to be developed.

After my clearance came through I ran the manufacturing work for the development model. We reported to Mike Maguire (Fig. 2.7) every morning. It was a very open meeting. He was not protective of anyone. He wanted the facts; for example, if something was wrong or was delayed on the floor he would say to me, "I want to know who held that up." In one particular case Quality Engineering was holding the work up. Gerry Braddon, who was in charge of Quality, was a very politically sensitive guy and he was trying to protect his organization. I was a little intimidated by him. Mike got me to tell him that his quality organization was creating a problem for us on the floor. I remember that very well.

The program was very complicated. It had a huge number of moving parts and still had an unbelievable record of success. Mike Maguire did a fantastic job as the general manager and the program was so successful. He was a very fair-minded guy. I never felt uncomfortable with him even if he had a very strong personality. I gave him the right information and let it lay there. Many times I went to him with my hat in my hand, but he didn't kill us. He didn't eat us up but ask, "What are you going to do to fix that problem?" I thought he was a very good leader.

Arnie Wallace's interpersonal skills weren't that great (Chapter 8). He had a lot of initiative to get the job done and pushed to correct problems. Many problems he figured out the solutions himself. You didn't like the way he was telling you things but he was right most of the time.

Once we were working on a job called "Letterpress" managed by Bob Landsman (Fig. 2.7). Bob Jones (Fig. 4.4) was director of Engineering at that time. Jones had a hobby of building cannons, and he used a lot of adhesives such as locktite so that screws were not necessary. These were thin film adhesives. We had this big casting into which the hardware was placed. Jones was insistent that we use adhesives to bond this big cylinder onto the structure. We were all kind of doubtful that this stuff would work because this was a big heavy structure and it had overturning forces, but we went ahead and bonded this cylinder anyway and he was happy. The adhesive cured and then a day later we went into the shop and there was the cylinder lying on the floor separated from the casting. You should have seen the expression on Jones's face; we didn't want to say anything.

BILL SCHILLING'S STORY†

I remember being back in the office when I got a call from Speedring in Detroit explaining that they had a thunderstorm with lightning and all the power in their building was killed. They had a large beryllium roller on the lathe and the tool created a big gouge into it as the machine stopped. So they asked us, "What do we do?" Ralph Sisk, the manager of Purchasing, told us to get out there quick. It was my order so I went the next day with big Walt Pleva. Walt was in the Manufacturing department and looked like a football tackle. We went there and looked at the part and assessed it. That piece was worth about 50 grand at the time and the question was whose responsibility was it. It was an "act of God." It was not repairable. We had to scramble to get another billet from Brush Wellman, the beryllium supplier. We went back to our customer and we wound up splitting the cost with Speedring as an "act of God." There was no insurance. We told old man Schiller, Speedring's owner, to put some protection in the building. And they did.

Once I was traveling with Oscar Berendsohn (Fig. 13.1), our metallurgist, and we got to LaGuardia airport after the 1970s highjackings. We were going to Union Carbide in Indianapolis and Oscar had a vial of liquid crystals that he was carrying out there. The purpose of the trip was to discuss Oscar's attempt to come up with a nondestructive test to verify the viability of electronic circuit boards by painting them with the liquid crystal and watching for color changes. If the circuit board failed due to a high increase in its temperature, the color changed. If the color did not change, then the board was okay and not harmed by the liquid. This test was not implemented because it was not reliable enough.

We went through the security x-ray and nothing showed up, but the airport personnel found the vial and they called him aside and started questioning him. Oscar got flustered while he tried to explain in broken English with his German accent what the vial contained. I finally had to go over and show company paperwork to explain that it was research and the liquid crystals wouldn't show. We finally got through, and poor Oscar he was so embarrassed. I loved Oscar and he was a great guy.

TESTING

Testing of the Hexagon subassemblies and the TCA were huge tasks that required preparation. The proposal had defined the major tests that would impact the design of the building that we would be situated in, as well as the incorporation of the very large equipment required for testing. Our company

†Bill Schilling was a buyer in the Purchasing department.

was required to supply the land, bricks, and mortar, and the customer would pay for the large equipment within. This included the vacuum chambers, vibration testing equipment, acoustic chamber, etc.

Jack Cunniff's Story[‡]

Hexagon, with its grand ambitions and dramatic complexity, required comparably grand capital facilities to support developmental testing of its many ambitious components and subsystems, as well as final testing to assure readiness for launch. There was broad and intense awareness that failure of Hexagon to work on-orbit would be a calamity. The Test department was tasked with development and operation of facilities capable of ensuring success for the grand venture.

Simulation of the space environment in which Hexagon would function was the dominating task assigned to the Test department. Space is far from a passive environment. Any object in an Earth orbit is cycled every 90 min through an exaggerated day–night experience. During the brief 'day,' the vehicle is exposed to intense solar irradiation while unprotected by any shielding atmosphere. During the 'night,' half of the vehicle is exposed to a view of cold black space.

Space is a harsh environment without twilight; an Earth satellite is either in the sun or in the dark. Transition is rapid and carries with it violent thermal shock. A vehicle in an Earth orbit when in sunlight experiences a continuum current of irradiating photons over the complete range from lethargic (only relatively) infrared photons that heat it, to energetic ultraviolet photons that erode the surfaces they strike. Spacecraft are made as reflective as possible to deflect these photons.

When in darkness behind Earth, a vehicle faces two exposures simultaneously. The underside, which faces Earth, views a surface at a temperature of roughly 300 Kelvin (K); that is, approximately the average temperature of Earth. The side facing away sees deep space where the temperature is slightly above 3 K (3 degrees above absolute zero). That is hardly any heat at all and causes space to act as a thermal sink where, because of the absence of a shielding atmosphere, every object or part of any object that faces space radiates thermal photons to it in proportion to the difference in the temperature of the object and of space but in an exaggerating way. The photon stream is proportional to the temperature difference to the fourth power.

Clovis Prendergast ran the Test department like a King of the Frank—mess up and he would behead you. It seemed that way. He was direct. He could be

[‡]Jack Cunniff was a senior member of the Test department.

trusted. Men working for him respected him. But he had limits. His reputation within his department was much better than it was outside the department, a thing that his team regretted and resented. During the intense period in which Hexagon work was pursued through three daily shifts, Clovis was unfailingly straight with his team and his team gave him undiluted loyalty.

But Clovis found Mike Maguire (Fig. 2.7) a tough boss. Mike's commanding intelligence and his ability to absorb detail intimidated many robust men. Clovis's position began to weaken when Mike instituted a daily stand-up meeting that began at 8:00 a.m. Directors and department managers were required to report on accomplishments against schedule, significant unplanned events, setbacks if any, and provide plans to advance the schedule. You had to stand up and quickly report in as a few words as possible. Then back to your department to resume your leadership.

Clovis unfortunately found starting his workday difficult. Consequently, and unwisely, he would go directly from the parking lot to Mike's meeting without first visiting his department's area in the high-bay where fabrication and testing of large vacuum chambers was proceeding around the clock. Mike, however, would have begun around 7:40 a.m. to phone the individual floor work areas to learn directly what had been accomplished on the third shift. He would then start his meeting knowing more than Clovis about status and would scold and humiliate Clovis for his gaps in knowledge.

When Clovis's loyal troops began to understand what was happening, our response to rescue Clovis was to ensure that every Test department man on the floor understood that no telephone that rang between 7:40 a.m. and 8:00 a.m. was to be answered. Mike's response to that, beyond scolding Clovis for the telephone unresponsiveness in his department, was to assign a retired Air Force Colonel named Harvey Henderson to be his morning bird dog and to snoop on the factory floor.

When Clovis's loyalists recognized what Harvey was doing by wandering through and asking questions, they instructed the floor people to answer every one of Harvey's questions with scrupulous accuracy exactly as the question had been phrased. They were confident that the questions as posed would reflect incomplete understanding of the detail nature of the work and would miss central things. The answers Harvey carried back to Mike bore flaws that Mike associated correctly with Harvey, so he caught Mike's wrath also. That ploy was only a palliative. It ended when Mike summarily fired Clovis. Clovis's successor was Charley Bryant, who eventually became in charge of servicing rework of our parts in the field at Lockheed.

Thermal Vacuum Chambers

There were quite a few thermal vacuum chambers, but two huge ones were required to test the whole system and they were our dominating concern. Their massive intricacy consumed us. The test object was intriguing. The TCA looked like a Siamese pair of transit-mix trucks joined at the hip. The TCA had to work for the simulated environments or calamity would befall.

The person mainly involved in the design and fabrication of the chambers was Bernie O'Neill. The largest was called chamber A and was a sphere 50 feet in diameter. The chamber was cryogenically pumped to enable achieving a vacuum level of 10E-6 torr or better. A simulated take-up was contained in a vestibule such that the test film could be removed for processing without breaking vacuum in the main body of the chamber.

The chambers were made of stainless steel. In chamber A, the day–night, sun-side/dark-side environment was simulated with infrared heaters to supply apparent solar photons and through shrouds that lined the chamber and through which liquid nitrogen, and for some tests, liquid helium, was pumped to provide thermal sinks into which the test object could radiate thermal photons.

Six 22-inch aperture collimators (large high-quality telescopes) were mounted such that three would be visible to each camera as the cameras rotated, representing scan angles of −45, 0 and +55 degrees. Each collimator projected a complex target containing a tri-bar target, resolution targets, registration marks, and a smear target. The targets were projected into the chamber through port windows, which were actually lenses designed to exhibit zero power when deformed by the pressure gradient across them.

The test article was connected to chamber A via a very large array of electrical cables to a sensor system test console (SSTC) located in a room about 50 feet from the vacuum chamber. Pete Cavanaugh was one of those responsible for the design of the SSTC. The SSTC was used for command and control; it received test data from the test article. The data was displayed in real time so that the test director and test conductors could ensure proper operation. The data was also recorded for post-test analysis. Tests were performed simulating worst-case hot and cold orbits.

Mike Mazaika remembered that there had been disagreement with the customer about the need for a big test chamber at the Perkin-Elmer facility. The chamber was for atmospheric testing of the camera system, seeing if it could work properly in vacuum. Our company thought it was required, but not everyone did. Mazaika explains, If we didn't have chamber B, would we ever be able to separate the variables if an anomaly occurred? I had the job of trying to justify the use of chamber B to the customer [CIA] and the Air Force. There was a power play with the Air Force and Lockheed trying to enlarge their part of the job, so they wanted to do all the final testing. No way was the CIA going to sign off on that. It became such an argument that the President's Science Advisory Committee, which normally only dealt with high-level intelligence advice, was also brought in for their opinion. Ultimately we did prevail and chamber B was used once, then never again for the program. But we wound up doing all the final testing in our chamber A.

BOB FINK'S STORY[§]

I started work in March of 1968 as an engineer in the thermal vacuum test section of what was then the Product Assurance Directorate. In August 1968 I started actually working on the installation and acceptance testing of chamber D with its solar simulator. Chamber D was used to perform optical testing of the optical bar prior to its installation into the TCA. The chamber was a cylinder 10 feet in diameter and 20 feet long.

Chamber D was supplied, I think, by Hi-Vac Industries. The Test department was tasked to operate the chambers; the procurement was actually handled by Clovis Prendergast's section. I was also involved with the installation and acceptance of chamber A, the 50- foot behemoth. A consultant was hired to assess the safety issues reportedly said that if the chamber developed a crack while under vacuum the resulting implosion would destroy the building—it didn't and it wouldn't.

Other than working seven days a week, the work was relatively uneventful. I was down in the bowels of the organization and was blissfully unaware of the turf battles among the higher-ups. I do recall that one Sunday afternoon while I was operating chamber D during its acceptance test but who should walk in but Mr. and Mrs. Richard Perkin with their neighbors. This was the co-founder of the company. Dick was just showing off his newest toys.

Once the facilities were accepted and the real work began, it was a 24-hour 7-day a week operation with no days off. The area was controlled, bunny suits were mandatory, and the players were told by the color of their caps—blue for test, yellow for quality, white for engineering support, green for everyone else, and most important, red for department managers and above. Getting a red hat was somewhat akin to being made a cardinal in the church. Although there were many departments involved in the test, Operations, Systems, Electrical, Mechanical, Quality, Reliability, Program Office, etc., the workers were really divided into two groups—those of us who did the work (the drones) and those who told us what to do (the queens). The drones developed a camaraderie that extended across organizational lines. This was especially true of the graveyard shift.

[§]Bob Fink was an engineer for many years in the Test department and had a keen sense of wit.

I operated the chamber: data was actually quite hard to get as the alignment of the interferometer was touchy and often bumped into the optical bar, causing vacuum problems requiring that the bar to be realigned then another vacuum pump down. The data was in the form of interferograms taken with a Hasselblad camera and Polaroid film that was then enlarged, eyeballed, and evaluated.

Vince Daniello's Story¶

When I first got to work on the program I was a test conductor on the test teams. I did system integration and test of the subassemblies onto the TCA, first on a simulator that went into chamber B. Subassemblies were tested first. I then got involved with system testing and I worked closely with the System Engineering department. Tests included film speed and film tracking from one end of the system to the other.

As the program grew reached the production stage, we eliminated the simulator tests in chamber B and we went right into integrating the TCA into chamber A. We did use chamber B on the first TCA simulator and the D model. In chamber B we used collimators to get resolution and effects of smear.

My first boss was Clovis Prendegast then Herbbie Weisblatt, all under Charley Bryant. Charley Bryant was funny. We had to call him up every night around 10 o'clock if we were on that shift to give him a report on the testing so he could be ready for Mike Maguire's early morning meeting. I give him a lot of credit because he was one of the few managers who went out of his way to find out what was going on in the evening. We rotated days and nights. I was famous for forgetting to make the call. I used to get hell all the time for not calling. What I did was to make a nice logbook for the turnover, then put a copy on his desk.

Charley was also very famous for coming into the place in the middle of the night to check up on us, so we used to tell the guard to call us if he came in. You don't know how many times we had to scramble out of the ready room from coffee breaks more than 10 min long. He used to come around the corner to the midsection wearing his red hat like all the upper managers and my guys there used to yell out to him, "Hey Charley, you thought you'd catch us in the break room."

Some people didn't get along with him, but my test team and I did very well under Charley. He was fair and he took good care of us. We had nothing to fear. He knew we had a good test team.

¶Vince Daniello was a longtime Test department employee.

We got along very well with the guys out there at Lockheed. We were working around the clock and we sometimes slept on the couch in Jim Lewis' office. I thought the guys on the West Coast did a fantastic job too, especially when Frank Harrigan took over.

One of the things that impressed me about upper management occurred one time we were downstairs and trying to integrate the forward path that went up into that little ante chamber in chamber A (chamber A was the huge spherical thermal-vacuum chamber that was used to test the entire system). The ante chamber could be closed off from chamber A so we could take some film out of the take-up without bringing the whole chamber back to ambient pressure and temperature. We had a lot of trouble getting the ante chamber opened up to retrieve the film.

Charlie Karatzas, director of Engineering, came in after the 4 p.m. meeting one day to see what was going on and he spent the whole second shift with us to try to open it up. He had his sleeves rolled up and he worked with the whole test team that night. I was the test conductor. Boy was I impressed, here is the head of Engineering working with us. It never did work out. He finally pulled the plug and said, "That's it, enough of this, it won't work." He said, "Let's shut down chamber A and go right to the take-up in front of the midsection; we'll retrieve the film that way." Charlie was quite a man, he treated everybody well. He was a hell of a guy.

Ed Belles was a roll-up-your-sleeves kind of guy. He had great rapport with people and told things how they really were. When we were working on a study for a related program, Ed was the program manager and he made me deputy program manager. We were working all sorts of hours and weekends. One day I complained, "You know, I have a load of lumber in my yard to rebuild my deck and I can't get to it, I'm always here." Ed replied, "This coming Saturday you are not going to work, you are taking off no matter what." Saturday at 8 a.m. my door bell rings: I was asleep—who was at the door but Ed wearing his carpenter's apron and his son John. His son was the same age as my kids, so the kids played together all day while Ed helped me put my deck up. Now there's a boss for you. Fantastic.

Many of the customer guys, like Berganini, used to come in at night to wander around and see what was going on. They would get involved to help troubleshoot and were very good. We used to give them jobs to do instead of them sitting around. We never treated them as "Oh, they were here to spy on us." We made them part of our team. They were great.

Bob Jones, director of Manufacturing and Test, was a big help too. One time we were walking around scaffolding in chamber B and he came in and thought it was dangerous, so he went back and got some of his mountain-climbing equipment like safety belts. He was worried about the safety of the test group.

Vibration Facility

Vibration was something else that required testing, and a facility was devoted to it. The vibration facility consisted of an arrangement of three random vibration shakers that could be ganged together. Each of the shakers was capable of providing 10,000 pounds of thrust to the test article. In order to properly phase the resulting vibrations, the three shakers were driven by a common amplifier. This arrangement was not sufficient to fully simulate launch-level vibrations for the entire vertically mounted TCA but was sufficient to screen out latent workmanship defects. The photographic system was not operated during the vibration tests. Each shaker by themselves was used to vibration test the individual major subassemblies such as the optical bar, supply, etc.

The noise and amount of motion of the assembly being tested was frightening. Bob Jones remembered how astonished and worried he was when the optical bars were shaken with their rotational axis vertical and the free end moved laterally at least +/– 6 inches during vibe.

An acoustic vibration facility was built as well. It was not large enough to fit the entire TCA into it. This facility ended up only being used to clean the individual large structures acoustically.

Photographic Resolution

Sam Palasciano had major responsibilities in the Test department and explains a special test area that was only accessible to several people due to its extremely sensitive work, namely evaluating the actual resolution accuracy of the photography.

> The facility contained a film processing and evaluation area. Several Kodak Versamat processors were used to process the test film from the chamber A tests. One rather unique piece of test equipment designed and built by us for this purpose was known as the auto-microdensitometer. This apparatus scanned the film, located the collimator targets, and read the data automatically.
>
> The processed film was examined for any unwanted markings that could be indicative of a latent defect such as film rubbing something. Throughout the program, weird electrostatic images would show up on the processed film. These were called tarantulas as they resembled a nebulous creature from beyond. These static marks also showed up on orbital imagery, always in the interframe space, never in imaged scenes. These were low-level exposures that were overridden by orbital imagery.
>
> Three of the engineers who evaluated the imagery were me, Dan Orlitski, and John Garrish. Dan was a retired Air Force officer and the

manager of the film lab. He lived in Massachusetts and usually flew his own little plane to the local airport and someone picked up and drove him up the hill and in 10 min was at his office. John was a systems engineer with a vast optical background.

FUNCTIONALITY TESTS

Basic functional tests were also conducted on some of the subassemblies, such as motors, caging, brakes, etc. There were some contentious discussions between engineers responsible for important parts of the system and senior management about the need for more testing. These would lead to complicated and costly test equipment and so a limited number of tests were performed.

QUALITY ASSURANCE

Each system went through the whole sequence of manufacturing, assembly, and test over a period of many months. It culminated in the chamber A tests that served as the definitive verification that the system was ready for shipment to Lockheed for incorporation into a launch-ready payload.

Throughout the program from design to production and test, the Quality Assurance Directorate was involved. There were quality control engineers and inspectors. The engineers had responsibilities throughout the process: signing off on all design drawings, making sure that the assemblies were fabricated to assembly drawings, and ensuring that tests were conducted per procedure. All parts were inspected and discrepancies reported to a Material Review Board where department engineers made the appropriate dispositions.

There are always design changes that need to be made to the design of various parts. These are usually made by the design engineers in the form of an engineering change request (ECR). After drawings were formally released, any changes had to be approved by an Engineering Change Board that was comprised of representatives of the major engineering departments. This board was led by Tom Miller who ran the meetings very strictly and that was a good thing. The requesting engineer had to justify the change and members of the board could ask any questions related to justification and cost worthiness.

SHIPPING

After all the parts were mechanically put together and electrically wired up, and optical and environmental testing done, the system was ready to be shipped to Lockheed in Sunnyvale, California. Our system was shipped in a

Fig. 14.1 Shipping container.

special huge air-conditioned temperature- and humidity-controlled shipping container (Fig. 14.1). Shown with the container is Bob Zarba, one of the vehicle managers.

Containers were trucked from our building in Danbury, Connecticut with a police escort to Bradley Field in Hartford, Connecticut and loaded aboard a C-5A (the largest cargo plane in the world) for its flight to Moffet Field in Sunnyvale, California adjacent to Lockheed where it would be integrated and tested in the entire vehicle. The vehicle was eventually shipped to the launch pad at Vandenberg Air Force Base in California. Support personnel always accompanied the vehicle on its flight to the West Coast.

Paul Convertito (Fig. 3.2) remembered that when they shipped the camera out to Lockheed in California the shipping container would come back full of cases of California wine and Coors beer, which couldn't be bought on the East Coast at that time.

Paul Brickmeier (Fig. 6.1) remembered that the caging mechanism on the first six flights had to be powered on during shipping, which was accomplished with 24 Sears Die Hard batteries. Every time the shipping container returned, the batteries were gone. It was odd that batteries could be stolen from this high-security project; it had to have been done by someone with access and clearance. Nobody wanted to make waves and start a criminal investigation; after all, the police could not be brought in because they did not have security clearance. These were just a few of the challenges associated with the secret environment in shipping the spy satellite.

Next we turn to how Hexagon operated in the field.

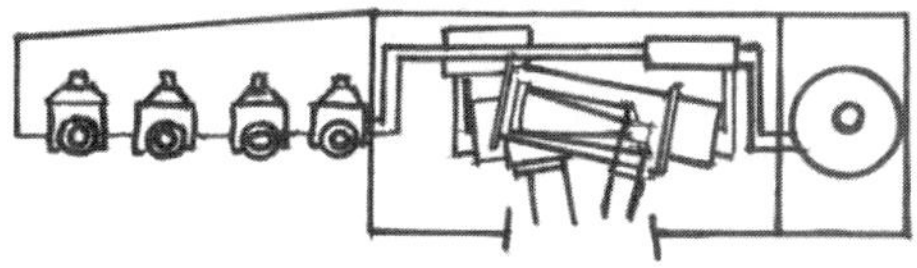

Chapter 15

FIELD OPERATIONS

FRANK HARRIGAN* and PHIL PRESSEL

In order to understand the overall organization of the program, take a look at Figs. 15.1–15.3. Figure 15.1 shows the satellite vehicle system responsibilities. Figure 15.2 shows subcontractor responsibilities. Figure 15.3 shows the satellite vehicle assembly and test flow.

In 1967 the Field Operations department began to form; Charlie Bryant was hired to establish and lead this function. Charlie was seasoned in the aerospace industry having been the base manager at Cape Canaveral as well as quality control manager and program office representative with General Electric's Missile & Space Division in Philadelphia. Charlie formed three operational organizations that reported to him. A small liaison group was located in-house in Danbury, Connecticut and headed by Aurie Health. The Midwest Field Operations (MWFO) was located within the McDonnell-Douglas facilities in St. Louis, which was responsible for the reentry vehicles. The largest group of Field Operations, the West Coast Field Operations (WCFO), was covertly embedded in the Lockheed Missile and Space Corporation (LMSC) (before Lockheed merged with the Martin Corporation) facility in Sunnyvale, California and was headed by Bill Cottrell.

Eventually Bill Cottrell left to take a position as the head of another important program. Sadly Bill Cottrell was assassinated while on assignment in Iran in the late 1970s. His skills and contributions to our company left a lasting positive legacy.

During 1968 and 1969 many covert reconnaissance programs had been terminated or scaled back, which created a recruitment boon for previously cleared personnel. The recruiting team of Charlie Bryant, Bill Cottrell, Frank Harrigan, Neimer Kaiser, and Lincoln ("Linc") Endelman all had past assignments and contacts in the black world; their goal was to attract known personnel with similar covert experience. As we hired employees, the team identified more candidates. As a result, staffing of WCFO and MWFO was accomplished with experienced and qualified personnel. Pete Krump,

Copyright © 2013 by the American Institute of Aeronautics and Astronautics, Inc., All rights reserved.
* Frank Harrigan was the Field Office manager.

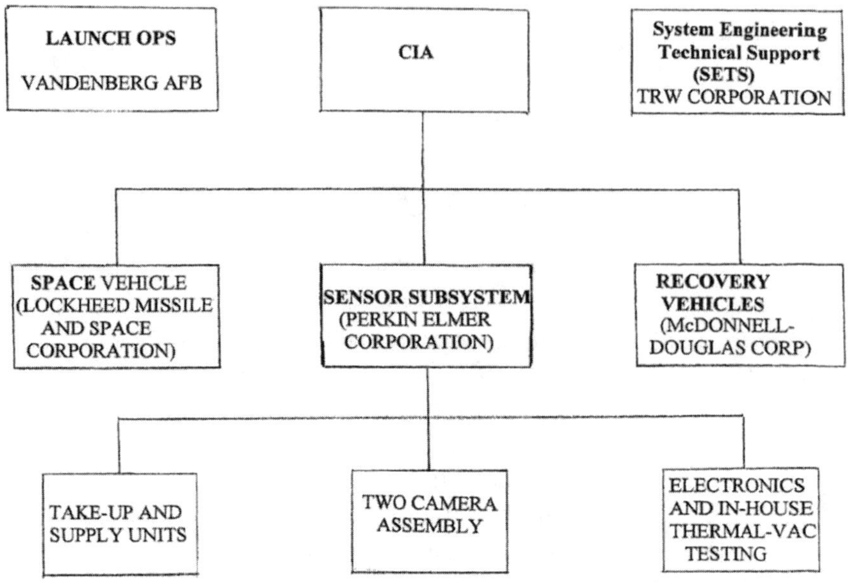

Fig. 15.1 Overall satellite vehicle system responsibilities.

Al Steck, Al Thompson, and Ken Dudschus were hired from General Electric; Dick Melberger and Don Squires came from Itek; Jim Dimas from LMSC; Larry Green, John Lenihan, Dominic Pascale, and many others transferred from the U-2 program within the company to the West Coast. Vic Abraham was the program general manager. Len Farkas was the chief systems engineer and eventually became director of the program. Bob Zarba, Bill McClafferty, and Pete Krump were vehicle managers. Frank Harrigan was the Field Office manager. George Manolis represented quality engineering. This was a formable nucleus for the Field Operations department.

Early flight successes resulted in operational mission lengths exceeding the design goal of 45 to 60 days. The first flight was in orbit 52 days, the second was in orbit for 40 days, the third flew for 68 days. Consequently, the launch rate was reduced; there were only seven launches between June 1971 and December 1973. Field Operations reorganized slightly to more efficiently accommodate this program change.

During the initial conceptual and development phase of the program, the CIA was responsible for the development of the sensor while the Air Force was responsible for spacecraft development and operations. Field Operations personnel established interfaces with all of the relevant contractors and facilities as well as with the Air Force operational units at Vandenberg and the satellite control facility. The TRW Corporation was awarded the contract

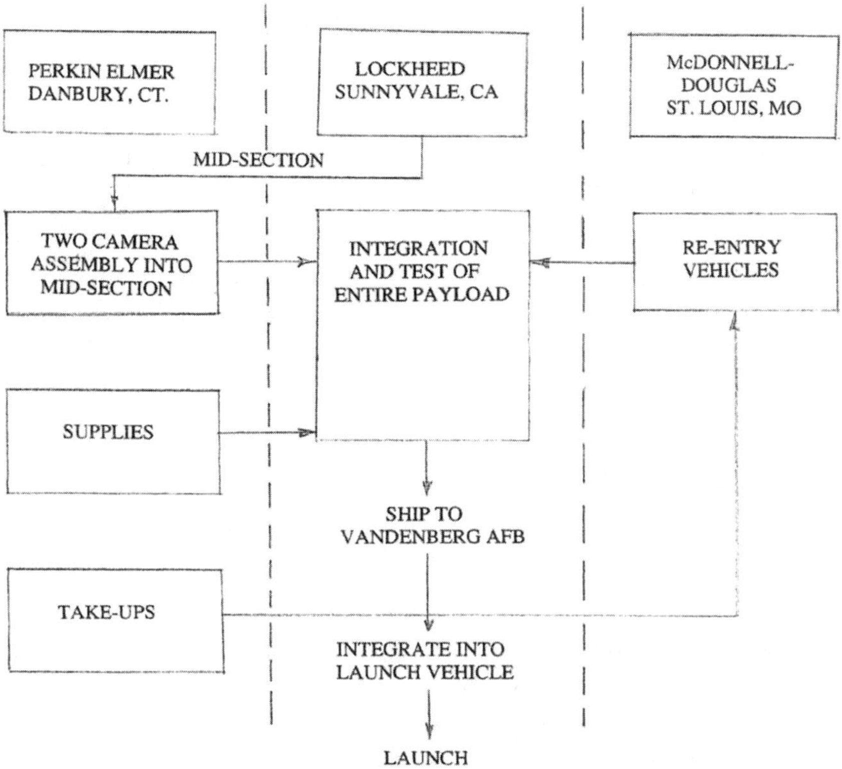

Fig. 15.2 Subcontractor responsibilities.

for the development of the flight operational software, which included mission planning as well as command and control.

These interfaces were developed and established with little difficulty as the personnel on both sides of the interface were experienced from similar programs and many knew their counterparts personally. The interfacing teams were very professional and extremely motivated.

During the initial phase of many earlier satellite programs there were an unacceptable number of early mission failures. So prior to the integration of the first Hexagon vehicle at Lockheed, a team was formed to investigate and examine the test flow starting at component acceptance through final assembly and ending at launch. The team consisted of representatives from all of the spacecraft contractors, the Aerospace Corporation, some independent aerospace contractors, the Air Force Test Wing at Vandenberg Air Force Base, and the customer. The findings and recommendations of the team were as follows:

- To incorporate integrated system testing in a near-flight simulated thermal vacuum environment.

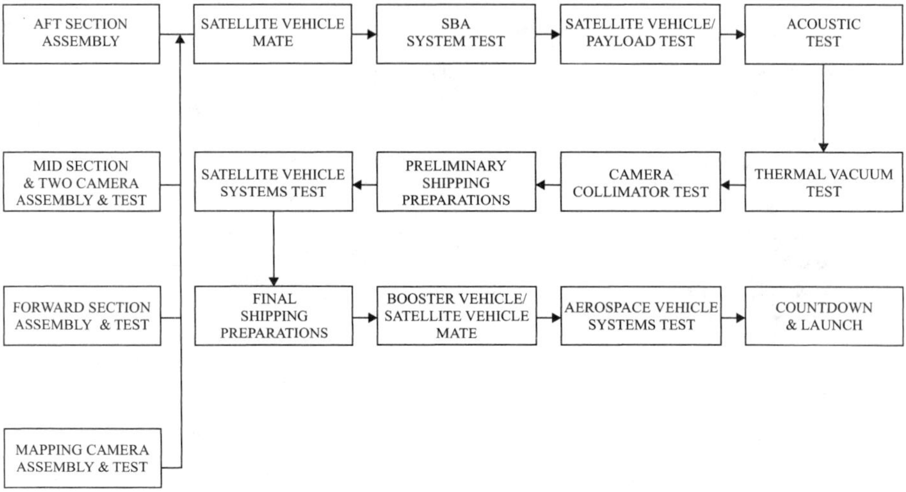

Fig. 15.3 Satellite vehicle assembly and test flow.

- To perform the final integration and acceptance at the integrating contractor's facility (Lockheed), then ship the spacecraft directly to the launch site. (Previously the final integration and acceptance was done at the missile assembly building at Vandenberg.)

Thus the concept of factory-to-pad was conceived. This last recommendation was quite bold and resulted in greatly reduced component failures and higher mission success rates.

The assembly and test flow shown in Fig. 15.3 represents the flow during the final phase of the program. The two-camera assembly (TCA), designed and fabricated by Perkin-Elmer in Danbury, was shipped from the East Coast to the West Coast in the section of the spacecraft's airframe called the *midsection*. The midsection, manufactured in and shipped from the West Coast, was used as a handling fixture for the sensitive and complex camera system.

The midsection containing the TCA was received at the Lockheed facility where it was subjected to a functional test similar to the factory system test. Meanwhile the spacecraft forward section was built up as the reentry vehicles, including the take-ups, film path, and mapping camera were integrated and functionally tested. The aft section, containing command and control, telemetry, attitude control, propulsion, electrical power and distribution, and thermal control subsystems, were integrated and tested. Finally the three sections were mated, forming the satellite vehicle (SV).

Dick Berg's Story[†]

I graduated with a degree in electrical engineering from the City College of New York in June 1966. My first job was with Kearfott, which soon became Singer General Precision of Little Falls, New Jersey.

One of my coworkers at General Precision was Marvin Krumper. Sometime in 1968 or 1969, Marvin went to work for Perkin-Elmer in the Optical Technology Division, in Danbury, Connecticut. After I received my masters degree in electrical engineering, also from the City College of New York, Marvin convinced me to work for him as a quality engineer at Perkin-Elmer in June 1969. I received my clearance in the fall of 1969.

My primary assignment was to get involved with any design and/or test concerns and ensure that the close-out rationale (resolution of the problem) was technically sound. My day-to-day world revolved around the generation and close-out of anomaly reports and Material Review Boards (MRBs).

The system was quite complicated, and I did my best to learn about as many aspects of it that I could. I very quickly found out the people who provided accurate information, and those who, while sounding very knowledgeable, didn't know what they were talking about. I even got to the point where I was teaching a class in the overall operation of the system. Marvin and the director of our department, Jerry, took great satisfaction in having a quality engineer teach a class on the system, as the Quality Engineering department wasn't held in the highest esteem.

I worked in Danbury until June, 1972 when I was granted a transfer to WCFO. My wife and I thought it would be an adventure to move to the West Coast. WCFO work was done primarily at Lockheed in Sunnyvale. Our assemblies were tested and integrated with the rest of the system. I started there as a data analysis and evaluation engineer and was soon promoted to supervisor of data analysis with four engineers in my group. Our work was done at Lockheed prior to launch, but other work was done at Vandenberg Air Force Base.

Our group did all of the evaluation and analysis of the servomechanism data from the various test phases to which the Hexagon system was subjected. The data that we evaluated was primarily in the form of strip-charts. There were multichannel representations of the diagnostics from the various servos in the system. The strip-charts were generated on a console referred to as the sensor subsystem test console (SSTC). The SSTC sent commands to the system to operate it in all of its various modes, either manually or by premade punched paper tapers. The SSTC also recorded diagnostic data on large magnetic reels and played back the data.

[†]Dick Berg evaluated test results.

During the course of testing, we had a data analysis engineer watching the data in real-time, and then we did a much more thorough post-test analysis. During the real-time data analysis, if we noted anything amiss we would tell the SSTC operator to send an emergency shutdown (ESD). The ESD command brought the system to a controlled stop in the shortest amount of time.

One vivid memory was of an incident that happened in the high-bay area when I was still in Danbury. The encoder engineer was performing a special test on an optical bar with a platen installed. They were not in the frame so a separate optical bar test set was used. Out of the frame, the optical bar by itself didn't care which direction it rotated in. In the frame, the optical bars rotated in opposite directions to counter balance and minimize momentum disturbances. The platen had cams and cam-followers that were designed to recock the shutter. The cams were designed to have a very rapid falloff at the end of the recocking cycle.

After a few minutes of running the optical bar, a loud screeching sound emanated from the platen area. The sound caused me to rush over to the test site. The encoder engineer immediately shut down the optical bar, but the damage had already been done. When I realized that the optical bar was being operated in the reverse direction with a platen installed, I told him what I thought the noise resulted from—the cam-follower bearings were forced to have a very rapid rise on the cam. They were meant to go *down* the slope, not up. The engineer then walked out of the area with his head hanging down. When the platen was removed from the optical bar, there were tiny ball bearings and shards from the bearing races of the cam-followers all over the place. The cams had been damaged from the remains of the cam-followers grinding against them. Unfortunately, the encoder engineer did not know about rotating the optical bar in the "wrong" direction.

Another memorable incident, also involving the platen, happened at WCFO. The platen had a set of flex cables that routed necessary electrical signals in and out of the assembly. The flex cables terminated in a connector located at the underside of the optical bar at one end-support. The wires were potted into the rear of the connector for stress relief. There was also a set of four light pipes that were routed in the same general area. In Danbury, some failures were noted in the wire terminations on this connector. Because the wires were potted into the rear of the connector with an opaque potting compound, a visual inspection wasn't feasible.

The folks in Danbury decided to have us x-ray the connector in place to determine if any questionable wire terminations existed. We performed the x-ray on one of the platens and all of the wires looked okay. Prior to x-raying the other platen, we did a test on the operation of the system as we had disconnected some connectors and removed some mechanical parts in order to perform the test. When the film was processed, it was noted that no information was present from all four light pipes. This was very puzzling, as it appeared that all four light pipes had simultaneously failed. While trying to figure out why the light pipes had failed, Danbury directed us to x-ray the potted connector on the other platen.

Because I'm a naturally curious person, I tried to think of some reason why all four light pipes had failed at the same time. We had previously seen failures of light pipes due to kinking or other mechanical damage, but never to all four at the same time. The only common element that I could think of was the x-raying.

As it was late in the day, I started calling light pipe manufacturers all over the West Coast. My question to them was "could x-rays degrade the transmission of light through a light pipe?" Most didn't really know, but one person said that if the intensity of the x-rays were high enough, it could affect the transmissibility of light through a light pipe having a specific type of construction. Because the x-raying on the other platen was going to be performed first thing in the morning, I immediately told the test manager to hold up doing the test.

First thing the next morning I called Danbury and expressed my concern. One of the managers there called me back and replied, "I've never heard of such a thing." However, they agreed to do a special test with the x-ray machine and a spare light pipe. After Danbury performed the test, the results were transmitted to the WCFO manager. He came over to me and said the special testing in Danbury showed that the X-rays had affected the light pipes to the point that they couldn't transmit any light. So I guess it was a good thing I brought the x-ray issue up.

Testing and Launch Preparations

System tests were performed on the vehicle demonstrating functional compatibility with all subsystems and the payload camera systems. The spacecraft was moved to a specially designed acoustic chamber and exposed to acoustic generated vibrations similar to the booster flight environment. After a comprehensive functional test, the spacecraft was subject to rigorous dynamic thermal vacuum testing that simulated both the ascent and orbital environments. All subsystems and camera systems were exercised. The collimator optical test was performed in another specially designed vacuum chamber that included two collimators where controlled images were used to establish and confirm camera focus, exposure and image motion settings.

Sam Palasciano remembered going out to the West Coast for a prelaunch briefing (Fig. 15.4).

Fig. 15.4 Phil Pressel, author (on left) with Sam Palasciano (middle) and Linc Endelman (who was based at WCFO, in charge of acquiring on-orbit data).

> It was chicken time. An Air Force general in charge of the meeting listened to everybody. They would all claim that they were ready to go, but almost everybody at the meeting had some problems that they were still working. Nobody wanted to be the first to say that they needed to delay the launch. This was serious stuff. All the program directors were there from the various contractors. The general had one of those FBI stares and asked, "Are you ready to sign off that you are actually ready to go?" And that's when the real truth came out. "Well I could really use a couple of days…" If it was a really serious problem, you'd get your extra days. There was no way he was going to launch the system and have a failure on his watch.

After tests were concluded, the shipping preparations were initiated. The flight film supply was loaded and the spacecraft was readied for shipment with the completion of a horizontal and vertical pre-ship test. Then the system was delivered directly to the launch site and mated to the Titan booster. Figure 15.5 shows the vehicle and the Titan 34D rocket on the launching pad.

The spacecraft was subjected to an equivalent vertical baseline test. Then it was prepared for flight including loading of expendables such as the fuel, the nitrogen gas into the pneumatic tanks for the pressurized film path, and whatever other gas or liquid components were required in other areas of the spacecraft. Countdown and launch followed. The spacecraft was injected into orbit downrange as operation was turned over to the satellite control facility.

Fig. 15.5 On the launch pad at Vandenberg.

Flight operations were conducted from the Satellite Control Center (STC) in Sunnyvale, operated by Air Force Systems Command and supported by many government and contractor organizations. The STC operated the satellite control facility, which consisted of the seven worldwide remote tracking stations (RTS) and the recovery control center (RCC) at Hickam AFB. The SCF tracking stations furnished the space–ground link that enabled the STC controllers to command, control, and direct readouts from the satellites. A technical advisor staff, consisting of representatives from all program contractors, was embedded in a closed area within the STC.

The National Reconnaissance Office (NRO) and other U.S. government agencies defined the targeting requirements and priorities. Knowing the orbital track and the predicted weather (cloud coverage) for the orbital swath, targets were selected depending on resolution, frequency of take, camera and sun angles.

During the preflight phase, previous missions' in-flight anomalies were reviewed and analyzed in support of in-house engineering. The upcoming photographic target requirements determined the mission plan defining the orbit parameters and mission duration. Then, working with the customer, the mission objectives were established.

A camera system engineering plan was published, the result of a series of meetings with Perkin-Elmer engineering, photo interpretation, and flight operations personnel, as well as with the customer and user representatives. The plan listed tests to be performed in-flight to optimize camera focus, exposure, and image motion settings. The plan also authorized the use of film necessary to perform these tests.

After Liftoff

After liftoff (Fig. 15.6), the camera system was subjected to tests before being released for operational use. First, the film path was operated to correct any film misalignment and loss of film tension that might have occurred during booster-powered flight. Proper film path pressurization and temperatures were verified. All telemetry sensors were checked as operational. Next, a precanned stereo camera operation was conducted; this test sequence had been run many times during ground testing and served as a baseline health check for the camera system. Rapid analysis of data permitted the camera flight operations personnel to release the camera system for actual photographic operations.

Fig. 15.6 Liftoff (left) and in-flight (right).

Fig. 15.7 The control room during a mission.

Figure 15.7 shows some of our personnel during a mission quiet-period. Ken Dudchus is at the left. Sitting are Len Farkas, technical director and chief systems engineer, reading a newspaper next to Vic Abraham, the program general manager. Behind Len is Pete Krump, the vehicle manager.

We often had to go out to Vandenberg for a launch, which we considered a fun activity. One time in the 1980s we went out for beers with some of our Air Force counterparts. We bought a bunch of Tsien Tao beers but the Air Force guys would not drink them because they were bottled in mainland China, which they considered an enemy.

During operations, occasionally engineering tests were performed on targets that were deployed at assigned sites in the southwestern part of the United States. These test parameters were recorded and delivered to the Field Operations post-flight analysis (PFA) personnel who supplemented a PFA team working at the processing facility at the end of each mission segment.

When the first of the four camera system take-ups in the reentry vehicles were full or near full, the film path was cut and cinched to the next inline take-up. This sequence was called *cut and wrap*. A constant velocity test was conducted to verify the successful cut-and-wrap operation before returning to operational photography.

When weather was favorable in the recovery area, the satellite was positioned to jettison the first reentry vehicle by rotating 180 degrees and pitching down. The reentry vehicle was released at the proper time, the satellite returning to flight orientation ready to receive another operational command. The reentry vehicle would then spin up, go through reentry, despin, deploy the parachutes, and shed the ablative shield. Successful airborne recovery was accomplished on all 19 satellites except for one reentry vehicle whose parachute did not properly deploy. That reentry vehicle sank before the nearby surface vessels could recover the unit (see Chapter 18).

As the program matured, new film types were introduced and short segments of experimental film were included in the flight film stack. These new and sometime exotic film types (color and infrared) were reserved for special engineering tests and sometimes for use on special targets. These film types presented quite a challenge for all flight operations personnel; the special film had to be readily available for targets and tests.

After the operational phase was concluded, a flight report was written and published. This classified report detailed the flight events and parameters. It served as the historical record for the performance of that camera system. Anomalies were recorded, including corrective action when required. The report also served as one of the inputs for another flight report published by the customer that also included an assessment of target satisfaction achievement and optical performance.

In the next chapter, we discuss the film.

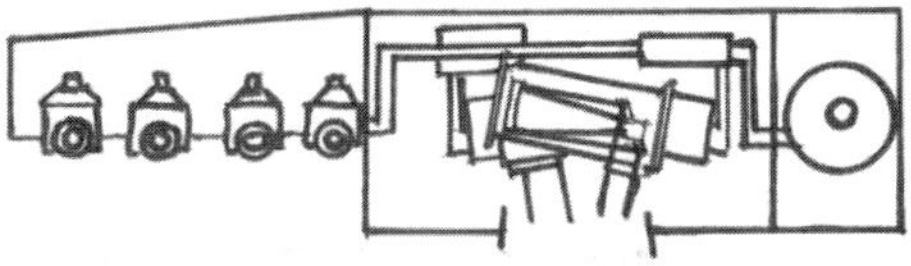

Chapter 16

FILM AND A DISAGREEMENT WITH THE CIA

SAM PALASCIANO*

A major Hexagon requirement for each mission was to photograph the entire Sino-Soviet landmass (Figs. 16.1–16.3). The nominal altitude over this latitude was approximately 100 miles. One degree of ground track was approximately 1 inch of film. These factors, coupled with the salient camera parameters drove the film capacity need to 100,000 feet for each of the two stereoscopic cameras. This converted to two rolls of film, each roughly 5 feet in diameter and weighing approximately 1000 pounds each. The Eastman Kodak Corporation supplied the film for the program.

FILM CHARACTERISTICS

To allow for adequate overlap between adjacent frames, a film format width of 6 inches was determined as necessary. A margin top and bottom was needed to allow for registration errors, tracking errors, and edge curl. A data block outside the imagery was needed as well. A final film width of 6.6 inches was used.

The film baseline was ultra thin based (UTB). This film was 0.002 inches thick. The emulsion was sensitive to black and white. Color film as well as false color infrared red (IR) was to be introduced later during Block III, the third contract with our customer.

The construction of the film was built around a thin 0.001-inch Estar (Kodakese for Mylar) base. The emulsion was coated on one side and was approximately 0.0006 inches thick. The backside of the film was also coated with an opaque layer of gelatin to balance the film to minimize warping due to hygroscopic effects. *Hygroscopic* means that the film for Hexagon

Copyright © 2013 by the American Institute of Aeronautics and Astronautics, Inc., All rights reserved.

*In this chapter Sam Palasciano (Fig. 15.4) discusses one of the most important features of the Hexagon system, namely the film. After the successful design of the system he was intimately involved in the analysis of film after exposure, including post-optical testing in thermal-vacuum chambers and after it was returned to Earth during a mission.

Fig. 16.1 Photograph taken in July 1980 of a Soviet Kiev-class carrier on the Black Sea in the former USSR.

required the environment's relative humidity to be between 45 and 55 percent—to keep the film from drying out. Both sides of the film were prone to the effects of relative humidity. When the humidity was low, the film would dry up; when the humidity was high, the film would absorb water. The relative humidity of the environment was especially critical for film flatness when the emulsion was deposited. If the film was maintained during all phases of assembly, test, and operation at a relative humidity of $45+/-5$ percent, then the film would remain as flat as when constructed. Thus, the relative humidity in our building was carefully maintained year round at this value.

If gradients of dryness were introduced into the film it would exhibit what would amount to a distortion, such as edge curl due to the variation in length across the web. This feature of the film was demonstrated early in the program by holding a piece of film by its ends and slitting the film piece along

Fig. 16.2 Photograph taken in September 1982 of a deep-space tracking antenna at Simferopol in the former USSR.

its length in 1-inch strips such that there were six strips, each 1-inch wide. The slitting was not carried through the entire length; the web could be suspended horizontally. It readily became apparent that the end strips, being dryer, were tauter than the center strips, which had retained more moisture (moisture was more readily lost through the edges of the film than perpendicular to the surface). The drier edges were physically smaller than the moister center; so, as one stretched the web horizontally, the tension was carried in the outer edges and the central slits sagged.

In the Hexagon design, the focal plane was an aerial image plane between two rollers without the film being backed against a flat surface. Keep in mind that the film was being transported up to a maximum velocity of more than 200 inches per second during photography. In order to meet resolution goals, film flatness was required to the micron level (1 micron = 0.00004 inches). Any gradient across the web would result in focus errors. If the edges of the

Fig. 16.3 Photograph taken in October 1982 of a typhoon-class submarine at Severodvinsk in the former USSR.

film dried out, then the profile of the film across the focal plane would not be a flat surface as the edges would be tighter than the center of the web—thus presenting a curved film plane to the aerial image.

PRESSURIZATION

Once the hygroscopic factors became understood and accepted, the decision to pressurize the film path was made. The supply was designed to be essentially a self-contained cocoon so that the integrity of the film could be maintained during extended periods of nonoperation during testing and on-orbit. This isolation was achieved by designing the supply assembly to be pressure and light tight, allowing only the leakage through a 4-inch diameter exit roller with a 0.005-inch gap between the outer film surface and the structure. This small gap was sufficient to maintain enough air within the supply cavity so that the relative humidity could remain stable. This tight gap restriction was often to cause grave concerns due to the possibility of contamination, splice thickness, and so on in terms of causing a blockage and catastrophic film damage.

PHOTONIC SENSOR

Two people who made major contributors to the understanding of mechanical properties of the film and film path interfaces were Bob Bosch and Don Cowles. Bob was most influential in such matters as moisture versus film flatness and was very analytical by nature yet extremely practical. Don Cowles, the patent holder of the twister, without which the entire design would not have been feasible, was a hands-on experimenter type (Fig. A.1). He devised a method of measuring film flatness while the film was being transported at 200 inches per second in a vacuum. His method utilized a *photonic* sensor, a noncontacting probe that was scanned across the film web and was sensitive to micron departures from the reference.

He incorporated this sensor as part of a film path simulator that was capable of sliding into a vacuum chamber, designated as chamber E in the test area of the company. Thousands of film tests under a wide variety of conditions were evaluated. The results of these tests revealed many subtle unsuspected performance-related factors, such as taper in the film thickness, hygroscopic state of the film, and tension gradients. This equipment became very important in support of anomalies encountered during in-flight operations.

Failure—the Film Breaks in Flight

In the early 1970s, the engineering (E) model was reconfigured as a one-camera film-path simulator that along with the chamber-E film-path simulator could be used to support (simulate) orbital problems. The E model had a simulated supply with all of the key elements in the film path, including a take-up assembly. The film path was not pressurized so that hands-on experiments could readily be performed. The optical bar was not operational and the film path was not light tight; therefore, no attempt was made to use the device for photographic tests.

During the second flight (mission 1202, launched January 20, 1972), three quarters of the way through the mission, one of the cameras suddenly experienced a film break. The broken end was wound into the take-up and subsequently recovered. The event was not accompanied with any untoward diagnostics that would be expected if the film were overtensioned or otherwise traumatized.

We experimented on the E model and were able to demonstrate that as the film was being wound on a supply stack at Kodak, the failure signature could be duplicated by depositing a single drop of water on the film. One drop welded the two layers together such that when despooling, the stuck area rotated around the diameter of the supply roll. When the leading web was sharply bent back on itself during despooling, it caused an edge rupture and then complete failure as the break propagated across the entire web. Almost no tension disturbances were noticeable during the failure.

Kodak subsequently added a drier to its air supply used during the winding of the supply rolls. No further such failures were encountered during the program. Note that it would take a tremendous tensile force to rip the film; but if the edge were to become nicked, it took almost no force to have the tear propagate across the web, thus parting the film. Bending the film back on itself caused a similar result.

Robert Kohler of the CIA offers his assessment of the problem:

> I can think of one technical issue that caused us some consternation. I think it was either the fourth or fifth mission where the film broke; the question was "why?" From your company's perspective they wanted it to be Kodak's fault; Kodak didn't have the same contract, but they wanted it to be the company's [Perkin-Elmer's] fault. I was in charge of that investigation. There were only two scenarios: one was that some sort of debris got into the film path and somehow poked a hole in the film and then ripped it; the other scenario was that when spooling the roll, somehow a drop of moisture got in there and welded two layers together and when the film came off it just ripped it apart.
>
> I remember dressing up in the bunny suit and going in when they got the film out of the bucket and cutting off that piece of film; I wasn't going to let anyone else touch it. Cutting off that piece of film was so we could do an analysis before it got processed or anything else. We didn't do any film processing. That was handled totally by the Kodak folks, but it was so important. We did a lot of investigation with microscopes and that sort of stuff. We finally concluded that it was some sort of debris. That conclusion came from how the tears actually occurred and running some examples of welding film together with moisture and it didn't tear it the same way. To my mind it was only an 80-percent conclusion. I wasn't 100 percent convinced, but the data pointed that way. It was kind of an interesting anomaly.

Conversely Dick Carricato (Fig. 8.7) of Perkin-Elmer had a different account, which is recounted in the following section.

Dick Carricato's Story

The phone rang at 2:30 in the morning. My wife Ginny and I were long past the stage where a phone call at 2:30 a.m. signaled a dire personal emergency. To her it meant that I'd either be rushing off to our company or rushing off to San Francisco. To me it meant that either there was a significant electronic/servo problem in one of the production systems undergoing test or an orbiting system was in trouble. Calls in the middle of the night were usually flight-associated because when it was dark in the

United States, it was daylight over the Sino-Soviet Bloc and the cameras were doing their thing.

This morning's call was from the team manning our war room. In language designed to be incomprehensible to an uninvited listener, they told me that the A-side had experienced an emergency shutdown (ESD) after the film had separated. In other words, the forward-looking camera had experienced an ESD after the film broke. Faxes of the quick-look data had been sent to the war room from the WCFO people, and the detailed flight data was being plotted and would be ready soon.

During the early years we had a crew manning the war room during every flight. The war room was the collection point of all flight data returning from the cameras. It was the place where missions were monitored and special operations were planned, and the place from which the analyses of the camera data were coordinated. Arriving there in the wee hours of this morning, my spirits were as gloomy as the roads I traveled to get there. This was the kind of problem from which there was no recovery. Once the film was severed in a camera, there was no way to rethread it on-orbit.

About one half of the total film supply (for both cameras) had been used up at the time of the failure. The remaining film on that one side, about one quarter of the cameras' total film load, could not be used. The loss of one quarter of the value of a very expensive mission could be blamed on us. It was a gloomy night indeed.

By the time I arrived at the war room, more data had arrived which revealed that the problem had probably started at the end of the camera operation before the one in which the film broke. During shutdown of the previous operation while the film was at relatively low velocity and heading toward zero, something snagged the film. The snag occurred somewhere between the supply and the input coarse tension sensor (CTS). For that matter, the CTS moved in the high-tension direction far enough to trigger an ESD. The system continued its deceleration and stopped without performing its usual rewind function. This was all normal for the ESD situation that existed. Actually, during the deceleration the snag cleared, the tension relaxed, and the ESD stimulus was removed, allowing the next startup to proceed normally. It was during this startup that the film actually came apart.

All of the technical people working on the Hexagon program knew about (and perhaps some were frightened by) the film exit vestibule (FEV) in the supply. This device was used to control the rate at which water vapor, originating in the film stack, could enter the film path during camera operation. The flow of gases from the supply was impeded by a relatively long, very narrow tunnel for the film and gases to travel through on their way to the rest of the camera. The characteristic that was thought by many to be associated with the current film break problem was the 0.005 clearance within the FEV through which the film had to pass. Normal launch vibrations could free any small pebble, fragment of film, or other contamination existing in the supply assembly, even those that might be trapped during cleaning operations. Then, in zero *g* they could float around until one found its way into that narrow passageway. The result would be very similar to the one we had just experienced.

My job was to analyze the data and try to make a determination of what had happened. I had to accomplish this without being prejudiced by those around me saying that the snag was in the FEV. By seven in the morning we had strip charts of all pertinent data. We had expanded scale strips for the final seconds of the run where the film snagged, as well as for the several seconds during the next startup where the film separated. The data included the summed error signal, angular velocity, and tension sensor position for all of the film path servos, plus the supply radius, the velocity command signal, and the looper position. Of course, we also had access to normal scale strips from all the other 100 or so data points within the cameras.

The most significant data were associated with the position of the input CTS. If we assume that the film was jammed in the FEV, it would mean that the speed of the film into the CTS was zero and the speed of the film exiting the CTS would be equal to that of the capstans. Knowing those two facts established how the CTS had to move during the snag. Comparing a few hand calculations with the CTS position data showed that the CTS was moving faster than that. How is that possible? In my mind there was only one conclusion—the film was also being pulled into the supply compartment. This would only happen if the exiting film web was stuck to the next layer on the film supply stack.

When the film traveled in the forward direction, the tension in the film reduced the torque required by the supply servo. If the film were suddenly jammed in the FEV, a step function of about 5 pound-feet additional torque would be required by the supply servo and this would be shown as a step in the servo summed error signal. The torque actually increased slowly and then even went higher than 5 pound-feet, indicating additional reverse torque acting on the servo. All of this confirmed (at least in my mind) the hypothesis that the film had stuck to itself within the supply stack.

The usual 7:30 a.m. meeting in the war room was not formally canceled that morning, it just faded away amid the hustle going on in there. The usual group of people was there, but this morning it was accompanied by many others. All of them truly wanting to help, but most of them were just adding to the bedlam. By 8 a.m. I had already made my first presentation to Mike Maguire. When I finished Mike suggested that I outline what I had just shown him on a few view-graphs, which I did. Shortly after that, a very stern Admiral Nimitz, the CEO of Perkin-Elmer, showed up in the war room asking what had happened. Maguire asked me to repeat for Chester Nimitz what I had just shown him.

The presentation was the same as I have given in the last few paragraphs here, except that it was aided by the actual data strips and a few hand-drawn view graphs. When I finished, the CEO looked at Maguire and asked, "Okay, Mike, what actually happened?"

Maguire made my day when he looked directly at the CEO, pointed a finger at me, and said, "Just–like–he–said!"

Evidently someone, most likely from Washington, had already pointed a finger at the FEV and us. Before this problem was fully resolved, there would be a lot of fingers pointed in our direction.

The next step was to get the analysis and simulation folks to simulate the situation where the film got snagged in the FEV and compare the resulting data with the data returned from the system on-orbit. We also had to look at the stuck film scenario and see if the data from that looked like the on-orbit data.

Meeting with the CIA

Before the morning was over we had been invited to a meeting at CIA headquarters in Langley, Virginia to present our findings concerning the failure. The meeting would be held early the following week, but I had a vacation scheduled to start at the end of current week. So I prepared my briefing materials for the CIA meeting and coordinated with several others who would attempt to get simulation results, look for any indication that film-sticking had ever occurred before, and shape my material into final form for the presentation.

I went to Hobe Sound, Florida (my first vacation in years), with a two-day break the following week to attend the meeting in Washington. This meeting was quite interesting. All of the players were there: Eastman Kodak, TRW (the government's technical support team), Lockheed, the CIA project team, and our company's team. All of these groups were presenting this day. They all sat and listened to each other's ideas and assessment of the failure. At least that is what I think went on—our team was asked to remain in the outer office until invited to present. When we finished presenting, we were asked to leave. It was quite obvious that Perkin-Elmer's point of view was not shared by all. We had no allies at that meeting—especially not the TRW team with whom I had spoken the previous week.

In a little over 3 weeks, with the mission completed, all of us met again. This second meeting was a bit different from the first. Perkin-Elmer was invited to participate in the entire meeting. We were even asked to present first, which made me wonder if we were being set up for something. But it wasn't that way at all. Our pitch, which was essentially the same as the first one, now had prettier view graphs, lots more data on film-sticking, and simulations that verified the sticking theory while eliminating the FEV as the cause of the problem. During the failure start-up we could trace the damaged piece of film that had stuck to the supply through the system until it reached the twister. It had managed to get through the FEV without trouble. It failed though when the film web, with a flap of material dangling from its center, encountered the twister air bars.

We also showed view graphs with pictures comparing the piece of film that had failed on-orbit and recently returned with a test piece that resulted from our attempts to duplicate the stick on-orbit. They were nearly identical. A single drop of water when wrapped into a test reel of film could cause the film to stick together as if glued. When despooled, the glued section would actually tear a piece out of the exiting web.

The second presenter was from TRW. I suddenly felt a lot better about them seeing our presentation first because it was if we had made their view graphs for them. Of course they had all the same kinds of simulation software that we had—the physics and geometry of the system don't change from the East Coast to the West Coast. But they sure didn't agree with us that first meeting.

The next presenter was from Kodak. We saw all the cleanliness controls used in the operations to make the supply reels. Kodak's bottom line was that nobody had ever seen the film stick to itself in the past, and there was no reason to believe it had happened this time. However we had performed a detailed search of flight and test data, which revealed that there had been several instances of small perturbations of the film coming off the supply that could have been caused by sticking. Not a cataclysmic one like we just experienced on-orbit, but small ones that could go unnoticed until specifically sought out.

At the conclusion of this meeting the various program managers met with the customer to discuss the performance fee that would be awarded for their product's performance on the mission. When Maguire came out of that meeting I asked if they had agreed with our assessment of the problem's cause. He shook his head, and I felt terrible. Then, with a smile, he said, "They gave us full fee!"

In the next section, Paul Brickmeier remembers other interactions with Kodak and the customer.

PAUL BRICKMEIER'S STORY

One of our engineers, Arnie Wallace, was tough as nails. But one day he had a heart attack. After returning to work whenever he was meeting with the

CIA and they got too aggressive, he would moan and grip his chest. Finally, Roy Burks from the CIA (Fig. 2.11) had to tell him to knock it off.

I met with Kodak where Mr. "Y" from the CIA attended. Kodak agreed to the film specification only if we agreed never to let Y talk. In all the years I went to Rochester, Kodak never took us to dinner, never took us to lunch. Suddenly, we get notification that we are going to go out one night. Kodak takes us to a restaurant; it must have been our people from Perkin-Elmer, the customer (CIA), Lockheed, and TRW. Stan Duffield, Kodak's chief engineer for Hexagon, stood up and made a little speech saying that Kodak had never agreed to a spec on the material [film] before. We would have to take it on faith that Kodak would comply because they never signed the spec with the government. But apparently now the heavens had opened: Kodak had changed their policy and would finally sign the spec. We were all stung because we were convinced they would never sign it. Stan then said, "But to get that signature requires one provision." This was in front of a big crowd of people, including Mr. Y from the CIA. Stan said that "That man Y will never darken our door again." Stan was serious! There was silence. Nobody knew what the hell to do. Then everybody laughed, but Y never showed up again and another government manager became the interface.

MORE PROBLEMS

Another mission failure event happened in one of the early systems during the 1973 time frame. I call it a take-up crash. For no apparent reason, the film off-spooled during flight and caused a catastrophic jam in the take-up. What we found, again using the E model and the film path simulator, was that the film thickness was not totally uniform. A wedge had developed in the take-up due to the winding of multiple layers of film slightly thicker on one edge than the other. A wedge of 0.025 inches across the width of the stack was sufficient to cause the film to climb the high side while being spooled onto the take-up, thus crashing into the stationary walls of the take-up. Upon discussion of this problem with Don Schoessler of Kodak it came out that the film was slit from large wide sheets that were not of perfectly uniform thickness, thus resulting in a slight wedge side to side. This wedge becomes exaggerated when multiple layers reinforced each other during take-up operation.

Two changes were incorporated to solve this problem. First, Kodak agreed to measure and control this thickness variation. Secondly, we found that changing the take-up builder roller design from an unwrapped roller to a 180-degree roller wrap made us less susceptible to the problem.

During the same time frame, we became aware that film flatness in the focal plane was also sensitive to this thickness wedge in the film. This time the wedge in question was in the supply rolls as received from the supplier. Wedge on the supply stack meant that the supply diameter was larger on one side of the film than it was on the other. As film was despooled from the stack, more film was being offloaded on the high side than on the lower side—causing a tension gradient side to side in the film. This gradient was such that an apparent tilt was evident in the focal plane. That tilt was reversed when the wedge was reversed. This directly degraded system performance in terms of focus nonuniformity across the field. This phenomenon was also corrected by Kodak tightening film thickness variations as Kodak wound a supply stack.

Film tracking through the torturous film path in the platen area was a constant concern that resulted in many hours and days, weeks and months of investigations in the simulators. Again it was found that film characteristics including wedged supply rolls were exacerbated. Dynamic friction between the film and rollers was another parameter of concern. Kodak attempted to remedy this problem by introducing grit to the backside of the film. We were unaware of this subtle change. Evaluation of several systems worth of flight film was noted for mysterious micro-blemishes in the photography. After detailed investigations, it was found that the tiny dots were caused by impressions of the grit in the backside of the film as pressure transferred the impressions into the emulsion.

Near the end of Block I (the first six systems) that was managed by the CIA directly, Bob Kohler of the agency ordered our participation in an effort to document all of these film anomalies into a Film Anomalies Handbook. Bob Bosch and I were sent to Camp Coolie (Westover Air Force Base in Springfield, Massachusetts) and ordered not to leave until all of these events were adequately chronicled.

About the time of the beginning of Block II (the next six systems), our immediate customer changed from the CIA to the Air Force. Several new film types were then introduced. The film was changed to ultra ultra thin based (UUTB). This black and white film was about 25 percent thinner than UTB and so we were able to increase the amount of film carried by 25 percent to 120,000 feet for each supply roll. Again we first determined system compatibility using our simulators. The film length was eventually increased to 150,000 feet.

Color film SO380 was also introduced. This film was thicker than the 0.002-inch black and white film; it was 0.0028 inches thick. Next came false color IR, about the same thickness as color. Both were found compatible with the system in terms of film flatness in the focal plane, tracking and stacking in the take-ups. These films necessitated a change in the in-flight filter in the optical bar. Block II therefore required the single filter capability to become an in-flight changeable filter design to enhance color operations. A two-filter interchangeable filter mechanism was installed into the field group of the optical bar to accommodate the new requirements. It proved to be quite successful.

Thus for whichever film was chosen for missions, the resolution improved as film types were developed. Even though the relationship between us and the Kodak Corporation was sometimes difficult, Kodak did come up with a good product that resulted in great success.

In the next chapter, electronics are discussed.

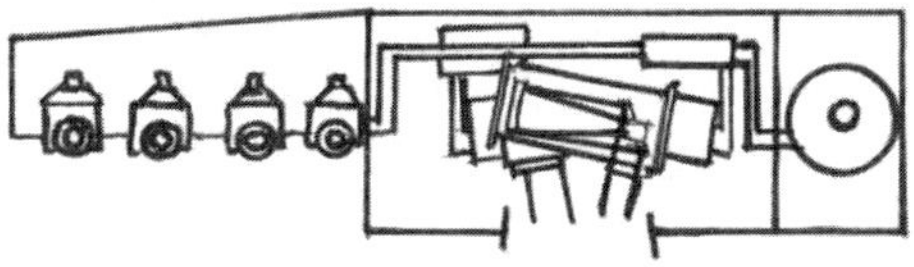

Chapter 17

THE ELECTRONICS

BOB WILLIAMSON AND WALTER FOWSKI

*BOB WILLIAMSON'S STORY**

I joined the Army in 1944 and was trained in The Signal Corps to repair their various radios, transmitters, and diesel generators. I had started making crystal radios by the age of 10 and repairing vacuum tube radios for neighbors. I was president of my high school's radio club.

I worked for Remington Rand Univac, where I was involved in the design of business computers and was a circuit designer. I also worked on electronic cryptographic equipment and became manager of computer development.

Perkin-Elmer approached me in the early 1960s (Fig. 17.1) with an offer to help them in the computer business. The job was to write the electronic portion of their proposals and then do the work upon receipt of the contracts.

CREATING THE DEPARTMENT

Mike Maguire (Fig. 2.7) had established a small electronic group and expected me to expand it. This effort expanded successfully and was the start of a functional electronics design support group. While doing this I had been put in for a clearance, so when the Hexagon program was awarded I was immediately designated as manager of this group, dubbed "electronics." I never had to go through the ordeal of the mushroom patch bullpen. The job was to develop/package the electronics identified in the proposal, which also included some off-the-shelf hardware.

Many of the electrical engineers that worked on the proposal were transferred into that rapidly growing group. My job was made easy by the outstanding quality of the engineers and my managers who trusted me to get the job done; they were a great family.

We newcomers were amazed and shocked by the size and scope of the program. We first set out to see what was required, including the technical

Copyright © 2013 by the American Institute of Aeronautics and Astronautics, Inc., All rights reserved.

*Bob Williamson managed the Electrical Engineering department during most of the program. He was a Perkin-Elmer employee when the company first bid on the Hexagon contract.

Fig. 17.1 Bob Williamson, longtime manager of the Electrical Engineering department.

specifications, the personnel available to do the work, the overall organization, the scheduling, and whatever else was needed to do the job. There was no off-the-shelf hardware. The design concepts were state of the art, with very high performance requirements.

HIRING THE VENDORS

A decision had been made during the proposal to use outside vendors to do the actual electronic packaging design and the manufacturing of the electronic boxes. Because our company had no electronic manufacturing capability, this was the only way to proceed. A well-known aerospace electronics manufacturing concern in Melbourne, Florida, named Rad-Inc (later known as the Harris Corporation) had already been singled out as a viable vendor. We were free to use others, including our own California-based Aerospace Division. There was no way all of vendor's employees working on Hexagon could be cleared for the job, so the work was done on an unclassified basis. It was sanitized in a way to make the final use of the product unknown. Little by little some key people were cleared and briefed to bring them onboard, as well to make the vendors a real part of the program.

Although the major concepts of the hardware were set forth in the proposal, after the award the customer came forth with a series of changes. The first included additional recovery vehicles. It was also necessary to resolve the electrical interfaces with the launch vehicle manufacturer

(Lockheed). This had not been addressed in the proposal. The proposal had shown no special hardware on our side of the interface, so we took that to mean it was up to the vehicle manufacturer to provide the power and signal electronics to directly accept our signals. This approach turned out to be unacceptable to Lockheed and as a result we were directed to add control interfaces (power, receivers, multiplexers, and electromagnetic control filters), a significant increase in our efforts.

We were grossly understaffed, so an all-out recruiting effort was required in all areas. The staffing was a nightmare, compounded by the need for program clearances that took an eternity. New employees while waiting for clearance were segregated in the mushroom patch. We had to break out meaningful unclassified work for them to do (not only to help get the job done, but to keep them from quitting); that was another nightmare. Morale in the group was low, and I often got calls from the Jolly Fisherman barkeeper reminding me that some of the staff members were behind in their luncheon bar bills. Some of our employees never got cleared and it was necessary to find other jobs for them in the company, or let them go.

Fred Marchianno, our administrator, set out to make a PERT schedule for the entire job. PERT charts identify specific activities and milestones, the sequence of the activities, the time required for them, and the critical path. The latter is the longest path in the project. The chart was updated as progress was made. It was so massive and complex that the biggest computers Fred could find crashed. It was then broken up into smaller chunks to make it work. The interrelationships between the mechanical and electrical design were very complex; each was dependent upon the other. This was compounded by the electrical engineers' worry that the mechanical design would move on so fast that they would be committed to electrical hardware that simply could not be built. This had actually happened to a few of us on other company jobs; when it did happen, it was sometimes a nearly impossible problem to solve. The project engineers for the major assemblies were all mechanical engineers; they were in control and under tremendous pressure to provide viable designs in a timely fashion.

There were also organizational issues. Traditionally the company project engineers had all the support people required assigned directly to them. They could often handpick support staff from the pool of personnel available. Management had decided that this program would use a functional organization and electrical engineers were therefore multiplexed between the many different mechanical assemblies, in accordance to function needed (such as motors, encoders, servos, etc.). This program, unlike past ones, required many such specialists. Nonspecialists were assigned to the Systems department, and that caused many conflicts.

Fig. 17.2 Bob Perliss.

While all these internal problems were being resolved, we were called upon to demonstrate to the customer and his advisors that we were proceeding correctly with the technical work and that we had a program plan which showed we knew what we had to do, how to do it, and were actually doing it. Within the first week we were required to make a presentation that described the electronic packaging effort. Fortunately during the proposal stage of the program, Bob Perliss, the electrical engineer in charge of electronic packaging (Fig. 17.2), had come up with a document that started the ball rolling. TRW, advisor to the CIA, wanted to take over from us a large portion of the program. They kept attempting at various times to make us look bad. After our pitch, a TRW representative started off his questioning with "What kind of connectors are you specifying?" This was a detail that was the least of our concerns at that time, but it set the tone for the rest of this meeting—and a sometimes contentious relationship. This was a disaster for me but eventually it got smoothed out.

One thing that was not a problem (at first) was money. Upper management kept after us to spend to get the vendors going and even to use our own small Aerospace Systems Division in California. This division appeared to be very talented, and very convincing, but none of us had any experience in working with them. We wondered what assignment could they be given that wouldn't ruin the job if they failed.

BUILDING THE BOX

We started out slowly by asking our folks in California to design a digital data acquisition system (DDAS). This was an electronics box mounted on the outside of the optical bar to transmit temperature data at various places. Unfortunately, the box generated more heat than the optical bar could tolerate. It was redesigned to turn on only when thermal data was requested.

The box also had to be gold-plated to get rid of the heat by radiation. It also served another purpose. It was located directly opposite the corrector window to act as a counterbalance weight.

Our electrical engineers became known for these gold-plated boxes. All other boxes were painted black. At this point we gained the confidence we needed to consider our Aerospace Division a very viable supplier. They didn't fail, and in fact, before the program ended they took over a very large portion of the work from the Florida vendor, who had become overwhelmed.

The DDAS had to communicate low-level signals through rotating slip rings on the optical bar. Slip rings operating in a vacuum don't work well; they tend to cold weld and can cause electrical noise. This led to a unique slip ring design using a mix of materials. The brushes were silver graphite doped with niobium diselenide, and the slip rings were coin silver.

Before subcontracting could be started, exact vendor specifications and statements of work were required for the different boxes and only a few could be done independently of the mechanical design. Electronics, motors, and encoders were required to support subassemblies such as the supply, take-ups, platen, looper, optical bar, and edge sensors.

We had to define high-reliability electrical components that could be used. High-reliability specifications were developed by the Reliability department, headed by Stan Karachuk, and negotiated with the vendors. New state-of-the-art components could not be used. Parts with a known history, usually five or more years of demonstrated reliability, were required. The engineers, subcontractors, and customer wanted to use microcircuits, but only a few were available at the time.

In the case of the encoders there were no solid-state light sources that met our criteria at that time. Therefore miniature incandescent filament bulbs were selected. They would fail under vibration unless selected to very close mechanical tolerances. The lamp manufacturers would not change their processes for us, which meant we had to select bulbs from standard manufacturing run units that had the right configurations. Bulbs that cost only a few cents commercially cost us around $400 a unit.

At one point I was so frustrated that I bought 100 Monsanto red light-emitting diodes (LEDs) that had just appeared on the market as a possible alternate. The electrical model shop was jammed with work, so I took them home and built a test rig in my workshop to see if there were any problems. I ran an electrical life test and within 10 h they started to deteriorate. Back to filament lamps—enough of the new state-of-the-art devices for a while.

Everything was interdependent. The mechanical design was evolving and so the electronic design evolved simultaneously. In the case of the servos, the question was who is responsible? We told the servo box designer, Marty Yellin, that he was, but then the Systems department said they were. Mike

Maguire's staff felt the System department was, and it was busy analyzing them. Meanwhile TRW also set up a servo group to determine the requirements (with more people than we had), and the box vendors wanted to get involved to help and to protect themselves. It all did work out well as Marty and his staff did a wonderful job designing the servos with excellent help from the Systems department (see Chapter 11).

The system was very complicated in that there were many sequential steps that needed to happen at precise times and to control film velocity profiles to extremely high accuracy requirements. A sequencer box controlled the steps. The proposal was based on an analog sequencer. It became obvious that it couldn't be done without adjustments that would constantly need tweaking, but this was not acceptable. Some of the earlier equipment the company had built had adjustments. It was a running joke that our engineers continued to tweak the servos even as the systems were shipped out the door.

Our rule for this program was no adjustments. So the sequencer was changed to a digital design. Stanley Anhouse, the section manager of digital design, was responsible for the design of the DDAS mentioned previously. During those days in the 1960s, most electronics were analog designs. Digital design was just being developed.

Box development was a long and expensive process. It took over a year per box in most cases, even with government priorities on parts procurements. Some boxes were small with as few as five parts. Most were much larger, with the sequencer exceeding 7800 parts, including 1272 microcircuits. An average box had about 1200 parts. The total part count of the electronics exceeded 45,000 and the total weight of the electronic boxes in each system was about 642 pounds. The system included some 42 different box types; one can see how both we and the vendors were overwhelmed with the size of the overall effort. Delivery schedules were never met. In many cases, boxes were shipped to support our production schedules before all the vendor work was complete. These then had to be returned for rework before the final mission could take place. The vendors were all proud to be able to say that a mission was never delayed because they failed to deliver.

Working in the Program Office

I eventually left the Electrical Engineering department for the Program Office. In 1978 I became involved in the negotiations for our third contract for six more cameras, with our customer at the time, the Air Force. They wanted to drive the cost down as the program was successful, had no major problems, and was a piece of cake. We felt that this would really drive up the risk like crazy. So Paul Petty, the division general manager, sent a team to California to negotiate with the Air Force. The head of the Air Force team was Clark Carlin. Our side was lead by Mike Mazaika. Mike was the director of the program at that time. Our bid not only included the funds required to do the job, it included funds to retain the skills required to maintain a viable organization that could respond to any and all problems. Due to the job security requirements, we could not just pull people in and out like a workshop operation. Those extra resources came under the category of our human resource plan, which was very expensive. We went in with an opening bid, which they quickly countered with an offer so low we were all stunned. The customer bids were, more or less, what they felt the hands-on labor was to build the product. Resolving these differences was long and tiring.

We worked for many weeks to find the specific areas of contention and then resolve them. The human resource plan was subject to hours of debate. It was October. As the negotiations dragged on, we approached the Thanksgiving holidays. Both the customer and our team were anxious to get home for the holidays to be with our families. It was the week before Thanksgiving when it became apparent that the customer was stalling, trying to get us to cave in so we could get home. Mazaika stunned them by telling them we were staying over. After so many weeks of restaurant food our taste buds longed for some really decent food; we wanted an old fashioned New England Thanksgiving dinner in Los Angeles. There we were at the Marina Del Rey Restaurant ordering all the traditional Thanksgiving foods. The restaurant made a stab at it, but it was a failure. From the turkey stuffing to the pumpkin pie, the tastes were all wrong. Most of us were disappointed.

When we reconvened the negotiations the next week, the attitudes had changed. The discussions were serious, and began to converge very quickly. In all, it was a total of five weeks, longer than any other negotiations, but it was successful for both us and the customer. The final contract was financially large for us and we could earn over 100 percent of incentive. At that juncture it was the largest contract that the company had ever signed.

WALTER FOWSKI'S STORY[†]

Jim Kiraly and I were aligning and testing the OAO-C spectrometer when the phone call came that would change the next decade of our working lives. After a security briefing we were told to report to our 77 Danbury Road facility the following morning, Jim going to the Test department and I to Electrical Design in what appeared to be a mere flip of the coin. Thus my ten

[†] Walter Fowski's relationship with Hexagon electronics began in late November 1966, while he was working for Perkin-Elmer on a NASA space telescope known as the Orbiting Astronomical Observatory-C (OAO-C).

Fig. 17.3 Walter Fowski, electrical engineer.

plus years of full time involvement with the Hexagon program began (Fig. 17.3).

When I showed up at my new home-away-from-home I encountered bedlam of the first degree. I reported to Bob Williamson and asked him what he wanted me to do. The response, as best I can remember as he darted from one fire to another, was "anything since everything needed to be done." I soon realized that there were some very talented people (such as Dick Carricato) working on servos, motors, encoders, and the sequencer, so I picked our interface with Lockheed and started to define what at one time was known as the data converter. Over time, the technical and contractual aspects of our interface with Lockheed became better defined, and many of the interface functions in the data converter were incorporated into the electronic boxes that came to be known as compartment electronics located in the aft compartment of the vehicle.

Soon we were involved in a marathon specification writing exercise as most of the detailed electronic box design and fabrication would be subcontracted to companies having more extensive capability and experience with space-qualified electronics. This was a bummer for those of us who wanted to do detailed design but it was necessary; with the staffing available, there was no way we could develop that capability in-house and still support the schedule required by our customer. On the bright side, I did feel I was contributing because I had experience working on government programs (Polaris A3 and Posiedon) when I worked at Lockheed Missile and Space Corporation (LMSC) while many of my coworkers came from the commercial world. In a matter of a few weeks we had pumped out design specifications for all of the electronic boxes required by the program. Most of the specifications initially went to Radiation Inc. in Melbourne, Florida for hardware design and development although some were redirected to our West Coast Aerospace Systems Division. I believe the logic behind that shift was to develop some in-house capability, albeit 3000 miles away, and avoid total program dependence on Radiation's performance. Other rumors for the shift were more political in nature, such as Radiation trying to contract directly with our customer instead of as our subcontractor, but I never bothered to confirm them.

Not having participated in the proposal stage on the program, I was not familiar with our customer (the CIA) or their method of working with a contractor. What I do remember quite vividly are the reviews conducted in one of the back rooms at 77 Danbury Road. Our customer, in the person of Bob Kohler, as well as being sharp as a tack, was ruthless when he heard stuff that didn't quite hold together. It was a lesson in humility and good training for the future. Our customer also was aided (I use that word with some reserve) by a team of technical helpers from TRW Corporation called SETS (System Engineering Technical Support) personnel. They were all over the place. It seemed like there were two SETS personnel watching and critiquing every one of us doing the work. It created a rather antagonistic relationship that continued until they departed a few years later. It is my recollection that the SETS team assisted the CIA during the preliminary design phase of the program, but was no longer needed when the Air Force took over the development and test phase of the program. The Air Force remained our customer until the end of the program in 1986.

In the late 1960s, the technology that we now take for granted did not exist. Transistors, diodes, resistors, capacitors, and relays were the common tools of our trade. Early integrated circuits (ICs) containing a few gates or a counter were available but large scale integration (LSI) was still a dream. Mission success (reliability) dictated the use of proven parts from Hi-Rel production lines with extensive burn-in to eliminate infant mortality. Part qualification was expensive and time consuming, resulting in pressure to use parts already qualified for existing military programs (meaning mature or semiobsolete technology).

The Fairchild uA709 became the operational amplifier of choice, and digital logic requirements were satisfied by the diode-transistor logic flatpacks. Packaging techniques included multilayer printed wiring boards (up to eight layers), welded modules, and EMI tight enclosures. Cabling (about 4 miles of wire) with bayonet-style connectors, like everything else, had to be class-100,000 clean and free of materials that outgassed. Due to the low orbit and short on-orbit life, no special action was taken for radiation-induced upset or degradation.

In general, identical electronic boxes were used in each camera; if slight differences were required, it was often handled via a jumper wire in a connecting cable connector. Almost every box contained its own power supply for secondary voltage generation.

To achieve stereoscopic photography, Hexagon's two cameras needed to be synchronized with each other. Synchronization was also desirable to control momentum disturbances caused by rotation of the optical bars, the platens, and the large film canisters (supply and take-ups). For these reasons, much of the control of both cameras was incorporated into a common, two-camera controller known as the sequencer. Other functions common to both cameras such as interfacing with Lockheed's spacecraft electronics (distributing power, commands, telemetry, etc.) were also incorporated into unique boxes that serviced both cameras. These common electronic assemblies were mounted on the aft bulkhead near the supply and were known as the compartment electronics.

In the spring of 1968 we moved to our new digs at Wooster Heights in Danbury and finally had room to breathe. Four engineers to a 12-by-12 foot cubicle seemed like a penthouse compared to 77 Danbury Road. Under direction of two very smart engineers, Bob Jones and Charlie Karatzas, the design and manufacturing engineering staff evolved into an enthusiastic pool of talent that could do just about anything; technical breakthroughs occurred not just once in awhile but day after day. Electrical Engineering department staffing continued to grow as additional Hexagon systems were procured and changes such as Block III IESD and S-cubed were authorized. By the late 1970s, staffing was required for new programs and it became necessary to replace talented people who were taking leadership positions elsewhere in the company. The department was a training center for future project leaders and managers, and headcount in the Electrical Engineering department exceeded 50 engineers during my tenure as department manager in the early 1980s. We were unstoppable!

To those of us at the bottom of the food chain it seemed as if upper management, in cahoots with our customer, worked overtime to make our lives more difficult. There were PERT charts to be updated (sometimes having little in common with the work being done), weekly activity reports to be prepared (requiring the use of a crystal ball since we had to turn them in well in advance of the period being reported), work-order/task-number tracking to ensure proper charging of time to the correct work breakdown structure element, and various other administrative deterrents to getting the work done. Over time I have come to realize that there were valid reasons for most of these reporting requirements but, back then, a less mature level of management understanding dominated my technical psyche.

Customer reviews often controlled our lives for lengthy periods of time. Major reviews exposed deficiencies in technical progress that often changed our work plans for an extended period. Monthly reviews had less of an impact but occurred more frequently. In both cases, preparing for reviews was challenging, with the program office often requiring multiple dry runs and the close out of an endless list of action items. Every once in a while I would get shipped out to the West Coast for a presentation at the Air Force that seemed very inefficient; two days shot for a half hour briefing to some colonel with a need to feel placated. But then, he did pay the bills (with our tax dollars).

If we weren't participating in reviews with our customer, we were attending reviews at Radiation Inc. or at our West Coast Aerospace Systems Division. A major part of our activity was to guide/monitor the development of the detailed electronics at our subcontractors via specifications and changes thereto. As the design matured and hardware fabrication began, subcontractor visits increased. Being on the road became a big part of our lives and the grist for many a tall tale.

Often guru Dick Carricato would be brought in to help us assess the problem and isolate the culprit. Dick had an amazing talent—the ability to take something very complex and look at it in very simple terms. He would go to the blackboard and block out a few simple transfer functions representing a very complex motor control loop, or whatever, and help everyone involved see the forest instead of each tree. I was forever in awe of his tutorial skills and attribute to him my acquisition of a brownboard that stayed with me for the rest of my career. I would use it to think, sketch, explain, and record. I found it helpful, and I owe it all watching Dick help diagnose Hexagon problems.

Over the life of the program, a few major electronic modifications were authorized and I was fortunate to have been chosen to be the electrical engineering project engineer on two. The changes were identified as the Block III IESD change and the-S cubed addition. Both were multimillion dollar contracts that took years to develop and deliver. Details of the S-cubed (a new stellar index pointing camera) are discussed in Appendix E.

SEQUENCER EMERGENCY SHUTDOWN CIRCUITRY

The Block III independent emergency shutdown (IESD) change had its beginnings in a system reliability analysis in the early 1970s. As I alluded to earlier, the sequencer provided an emergency shutdown capability that would bring both film paths to a safe stop if triggered by a looper position anomaly. Such an anomaly would be created if the average velocity of the fine path was different from that of the coarse path, a situation that was not good to say the least. For most system electronic component failures, the ESD capability would save one of the cameras (the one without the component failure) and allow subsequent operation in the monoscopic photographic mode. However, the failure mode effect and criticality analysis did identify numerous electronic component failure modes that would prevent a safe ESD or subsequent operation of either camera. In conjunction with someone from the CIA we began to explore what it would take to eliminate all single-point failure modes that would disable both cameras.

Given the complexity of the sequencer, modifying the design to be actively failure tolerant (active redundancy) was out of the question. Making it passively redundant was doable but we still needed to ensure that we could safely shutdown at least one camera under any condition of failure. To achieve this goal, the ESD circuitry in the sequencer needed to be independent of the rest of its electronics, hence we called it IESD. With this approach as a foundation, we next attacked making the IESD sensing system better to ensure we would have adequate warning of all possible failure modes before catastrophic damage was inevitable. We also needed a better technique for grabbing control of the system and bringing it to a safe stop as fast as possible.

Working with me on this endeavor were Gerry LeBlanc and Dick Whittlesey, often referred to as the Bobbsey twins. They would invent problems just as fast as they would solve them...never a dull moment. Then there was the primary consumer of all Wooster Heights classified computing power, Franz Holzer. He would run system simulations that would tell us what would happen if hypothesized failures occurred and various proposed IESD capability existed. Not only did he have to remodel the simulation to reflect the changed capability but the runs would take hours to complete. After numerous runs, we would go to Franz as a group to ensure we would survive his wrath. We could always tell when we were in trouble. His German accent would thicken, and he would terminate all conversation with one of his favorite sayings—so-and-so (usually a member of the Program Office) "needs a brain transplant." We finally came up with a winning approach: make the sequencer totally redundant.

The CIA gave the go-ahead and then turned it over to the Air Force for development and deployment. Major Dave Berganini was put in charge of monitoring the activity, a reasonable guy and easy to work with. I think he actually enjoyed getting away from the day-to-day administrative problems and politics of his job to mess around in the technical mud with the Bobbsey twins and me. He often joined us for reviews at Radiation Inc. where most of the detailed design and fabrication took place. What was once a 5000 component box, weighing 60 pounds, had become a 10,000+ component box, weighing 110 pounds.

Concluding Remarks

Recalling events, people, and technical details after thirty years without the benefit of notes has been a daunting task. Nothing above should be taken as the absolute truth; it is merely one man's perception of ancient history. What can be said with more conviction is the impact that the program, the challenges, and the people have had on our lives. Starting with only 5 years of experience, the opportunity for me to learn and grow was unlimited. Not only were my coworkers smart, they represented many different technical disciplines and all levels of experience. There was not only an electronic solution for any problem but perhaps a mechanical or optical solution as well. Hopefully, we chose the best solution most of the time. It was smart people—creative people—working together that made Hexagon such a success.

Given the Cold War and the conflict in Vietnam, we were fortunate to be participating in a program that was not building offensive weapons, defensive weapons, or deterrents to war as I had previously worked on at LMSC. Hopefully the product of our efforts provided the information that made the use of such weapons unnecessary and the world a safer place for our children to inherit.

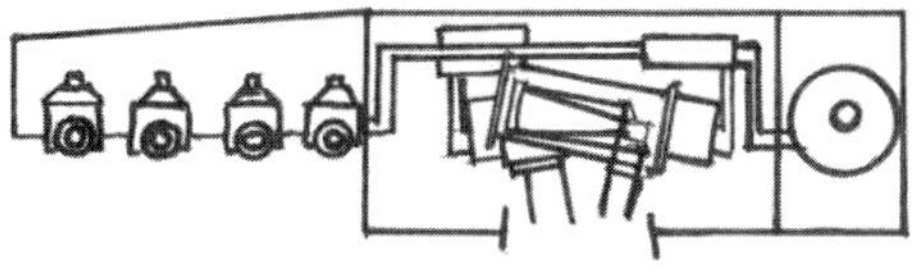

Chapter 18

Sunken Reentry Vehicle

Phil Pressel

Reentry vehicle/take-up recovery was typically completed via an in-air, over-the-ocean catch as the unit descended on a parachute. As the capsule descended, it was detected and tracked by tracking stations and by recovery ships. An Air Force C-130 recovery plane located the capsule by homing in on a radio beacon attached to the capsule. Once the capsule was sighted, the pilot conducted a fly-by to inspect the capsule and to make sure the parachute and nose cone had deployed properly.

The aircraft used a trapeze-like device with grappling hooks attached to the end of two long poles extending from the belly of the plane. The objective was to fly over the capsule and snag the parachute, then winch the parachute and capsule into the aircraft.

If the multiple-pass opportunity to catch was missed, the valuable package was designed to float on the surface for a prescribed time—allowing water recovery. Surface ships carrying helicopters and frogmen attempted to reach the capsule before it sank. The frogmen would jump in the water and attach a line to the capsule; then, it would be winched up into one of the helicopters. It was also possible to parachute pararescue men from a helicopter over the water where the capsule had come down. They would inflate a raft and secure the capsule to it.

First Launch

Before describing the recovery of the only reentry vehicle to sink during the lifetime of the program it is worth telling the story of the first launch of Hexagon.

Dick Carricato (Fig. 8.7) was one of the few lucky engineers sent to California for the first launch. He was excited but quickly realized that he was not going to see the launch live, but instead was going to be sent to an office that Perkin-Elmer had at Lockheed's complex located behind the Air Force Satellite Control Facility in Sunnyvale. The Satellite Control Facility was

Copyright © 2013 by the American Institute of Aeronautics and Astronautics, Inc., All rights reserved.

more popularly known as the "Blue Cube" because the building was a large cube sided with blue-tinted windows. "On launch day that area would be just about devoid of people," Carricato remembered. "However, there would be no lack of diagnostic data from the cameras. I would be able to do my thing; stare at roll after roll of data returning from the vehicle's data system and assess the health of the cameras." Maybe not the same as being at the launch site or in the Blue Cube control room, but still an exciting location for someone who had worked so hard on the project.

And Carricato had worked hard, sometimes 80-hour weeks, no family life, hardly any thoughts except Hexagon and getting to the first launch. As a special treat, the first launch was taking place on June 15, 1971, exactly six years to the day since Carricato had joined the company. "Although we couldn't see the launch from where we were, we could listen to it via an intercom system," Carricato remembered. "It appeared to proceed right on plan; the count-down, lift-off, a dog-leg maneuver [so that the rocket would not overfly some islands off the California coast], orbital injection, and all the booster and strap-on events worked just right to place our payload at the right altitude, at the right speed, and in the right direction!" Carricato recalled thinking that a lot of people had done their jobs very well indeed. But now it was Perkin-Elmer's turn. Could they do as well? They'd know in just a couple of hours.

This was the era before communications satellites. Most of the data generated within the spacecraft had to be recorded and then transmitted to a ground-based tracking site when the satellite came within range. However, an engineering pass could be designed to operate the cameras while passing over a tracking station. In that case, the data would be sent directly to the tracking station where it could be recorded and distributed immediately via land lines to the users. That was the mode the spacecraft was operating in during those first few hours of the mission.

Carricato was there looking at the raw data when it came down and instantly became alarmed. The data didn't look exactly right. Lots of little things were different, off by a little bit. Did it mean that something was failing? The camera system was operating, but Carricato had seen hundreds of startups during ground testing and the data coming down from the spacecraft in orbit didn't look like that.

Then it hit him. "What I had been observing for years were camera systems quite different from the one I was now watching come alive on-orbit," Carricato explained. "All the previous ones had been sitting on the ground operating under the influence of 1 *g*. Usually the film paths were centered before the start of testing, all of the tension sensors and loopers were manually positioned to their nominal positions, and the optical bars were pointed straight upward. These systems didn't even have a forward section with its four take-up assemblies. They had a take-up simulator that allowed easy film retrieval for processing at the completion of testing." In fact, the whole system had never been complete until it was integrated with the real take-ups in Lockheed's final assembly facility. "This was the first look any of us had ever had of the system working in its native environment."

"All of my concerns, and there weren't very many when one considers the vast number of things going on in these two cameras, could be explained by small changes due to zero *g* operation, and the fact that the loopers are caged for launch." Things like the tension sensor variations that had concerned Carricato were the result of the looper being driven to its nominal position by the drive capstans. "As the cameras operated, their electronic signatures started looking more and more like the ones I remembered from the test sequences at Perkin-Elmer. I started to relax when I realized that the cameras were functioning as well as the systems that had brought them to space." Carricato breathed a sigh of relief.

But of course the film was being exposed up in space. It still had to come back down to Earth in the reentry vehicles, and it had to be processed and assessed. This would all happen in about ten days.

First Recovery

"I was invited to the Satellite Control Facility for the first recovery of a capsule from Hexagon. There was nothing I could do to assist in this phase of the mission. I and the camera flight operations team could only observe as other groups performed their part of the mission," Carricato remembered. "Our part of the recovery mission had taken place a few hours before when the film on the first take-ups in RV-1 was secured, power and control was then transferred to the take-ups in RV-2, and the film was cinched onto the take-up reels." The final actions before recovery could take place now, included the sealing and separation of the film path as well as the cutting of the electrical connections. "All we could do was to wait. There was no video broadcast like the Apollo program had for their splashdowns. For that matter, we hoped that there wouldn't be a splashdown."

Instead, the plan was for the reentry vehicle to slow down and fall through the atmosphere, heating up. Once it had slowed down sufficiently, it would deploy a small drogue parachute to stabilize it and then a much larger parachute. All of this would happen several hundred miles northwest of Hawaii. A fleet of Air Force C-130 aircraft flying out of Hickham Air Force Base on Oahu (which had been attacked by Japanese fighter planes nearly three decades earlier) would move in on the descending space capsule. A C-130 aircraft trailing a cable strung between two poles sticking out its cargo bay would then line up on the parachute. If the pilot had timed everything right—and the pilots did a lot of practice to get everything right—grappling hooks hanging along the cable would snag onto the parachute lines and catch it. The C-130's crew would then winch the reentry vehicle with its precious film cargo into the cargo bay and the pilot would head back for Hickham.

Carricato didn't get to go to Hawaii for the recovery. He was still out in California. "Through the intercom within the Blue Cube we could hear the conversation between the pilot of the airplane and the tracking and control facility," he recounted. "This was a classified operation so no explicit reference to what was going on was allowed. A simple sheet of about 50 'events' was handed out to the group listening to the recovery operation." Carricato no longer remembered most of the events on the list, but he did remember that the pilot would say things like "event number 34" and they would then run their finger down their list to number 34 and see that it said "visual contact with the parachute." Then there was a sequence of events as the pilot maneuvered to make a fly-by inspection of the parachute before attempting the snatch. After the fly-by inspection the pilot reported, "event 24."

"On our list we read, 'parachute damaged, air recovery impossible.'" Carricato was disappointed. "All sorts of negative thoughts passed through my head. Would the capsule survive water landing? Would the film get wet? Would the water recovery team find it before it sank?"

They didn't have long to wait. "In a very few minutes we got word that the capsule was on the recovery vessel. Someone said that it was in the water for only three minutes before it was picked up! It was dry inside." This was one more example of several groups of people doing outstanding jobs. The spacecraft had to be pitched to a precise angle, the reentry vehicle had to be spun-up and released, the thrusters had to cause the correct velocity change, the heat shield had to perform, the parachute had to deploy, and all had to happen with great precision to make the "bucket"—their name for the part of the reentry vehicle that actually held the film—land within a small footprint near the Hawaiian Islands. "The only bad thing was that nobody could talk about it," Carricato said.

Carricato remembered something amusing upon his return east. "Shortly after returning to Danbury I was shown the back sports page of one of the popular tabloids." In large black font were the words, "A HEX HANGS OVER THE REDS." Was Cincinnati losing a lot of baseball games or was there another message in this headline?

Checking on what had happened to the parachute, Carricato learned that the part that was damaged was a special bubble at the top center of the parachute used for air recovery. The main part of the chute was too big to fit between the C-130's poles with the cable strung beneath, so the bubble was there for the pilot to snag onto. The main supporting element of the chute was fine, as was proved by the successful water landing and recovery.

And how did the film look? "It looked so good that everyone in the Perkin-Elmer plant where the cameras were made got to see the results of all their efforts. Hundreds of feet of enlarged prints of the film returned by Hexagon were tacked to the walls," Carricato remembered. These were all shots of the United States: particularly of the northeast coast, the workers' homes, and the buildings occupied by Perkin-Elmer along route 7. "It was viewed with pride by one more group that did an outstanding job."

Carricato had a lot of respect for the customer. "I must say that most of the guys I met from the CIA were impressive," he said. "They were smart, intelligent, dedicated guys who seemed to really believe in what they were doing."

Trouble at Sea

If water recovery didn't occur in the prescribed time, the reentry vehicle would be flooded and the unit would sink to the bottom in very deep water. During the first mission, launched on June 15, 1971, the parachute on the third reentry vehicle failed to deploy properly. The reentry vehicle struck the water with sufficient force to break the sealed volume, and the precious cargo sank.* This happened on July 10, 1971. The reentry vehicle carried approximately 1100 pounds of film. The recovery control group's mission was to effect aerial recovery of capsules ejected from orbiting satellites.

In an effort reminiscent of the Glomar Explorer being used to retrieve a Russian submarine, an effort to recover the film was made for this sunken reentry vehicle. Preparations for recovery included a deep-sea search for the payload; design and deployment of a special lifting claw; preparation for special handing at the film processing facility that included provisions for sea water immersion; and accommodation of any impaired ability to rotate the film roll on its supporting bearings. The effort to locate the payload was eventually successful with photographs revealing that the take-up structure and film roll had broken free of the reentry vehicle upon water impact. The film experts at Kodak believed that the images on the film would be retrievable at some distance in from the edge of the film because edge swelling in the seawater could provide an effective light and contamination shield.

This reentry vehicle contained approximately one quarter of the total amount of film for the mission and in particular contained pictures of an

*David Waltrop, "An Underwater Ice Station Zebra," *Quest: The History of Spaceflight Quarterly*, Vol. 19, No. 3, 2012.

important area of the world that was normally under clouds, but happened to be cloud-free during one photographic period. It was therefore important to retrieve this data.

The customer (the CIA) gave Perkin-Elmer the task of designing and fabricating a claw and of supporting the mission to retrieve the reentry vehicle. My colleague Len Molaskey,[†] the manager of the Mechanical Engineering department at that time, was given this assignment and became dedicated solely to this task, as were a designer and a manufacturing engineer. Nick DeFilippis, manager of the Mechanical Engineering department, attested to Lenny's capability. Nick went to Kodak once with Lenny. "He took me into this dark room, and he's working completely in the dark. The son of a gun knew exactly what he was doing. He had the ability to look in there and look at the different components and figure out how to take it out in the dark. He was very good at that."

The claw had to be designed to grab the reentry vehicle and haul it up to the surface of the ocean from a depth of 16,400 feet. It was necessary for the claw to be able to attach itself to the shape of the reentry vehicle and hold on delicately enough to prevent further damage. It was assumed that water penetrated the reentry vehicle structure, thus flooding the interior.

As the claw was to be operated remotely from a deep-sea submarine (what the Navy called a deep submergence vehicle), it required a mechanism to open and close the jaws of the claw. The structure of the claw had to be strong enough to support over a 1000-pound load during the lifting process (Fig. 18.1). There was no need for the claw to withstand the pressure of water at depth because it was an open structure with no sealed cavities.

The claw was designed, fabricated, and then shipped out to Hawaii (Fig. 18.2). Meanwhile the customer arranged with the Navy to make available the three rescue vehicles. The actual underwater retrieval was performed by the Trieste II Deep Submergence Vehicle-1 (DSV-1) that was stored, launched, and supported from the support ship White Sands. The vessel Apache was used to tow the White Sands (Fig. 18.3).

Len Molaskey flew to Honolulu in early October and soon went aboard the White Sands for its expedition into the middle of the Pacific Ocean. According to the *Quest* article referenced earlier, in September 1971 the weather was good and test dives were conducted over several weeks to practice retrieving a dummy bucket. The Navy had previously searched and continued to search for the exact spot where the capsule had sunk. They were aided by recovery data and the use of echo sounder profiler and sidescan-

[†] I give thanks to Len's son Richard Molaskey for providing me with some of the photographs included here and for additional information about his father's mission on this task. Richard also worked on the program, and his insight makes this chapter more interesting.

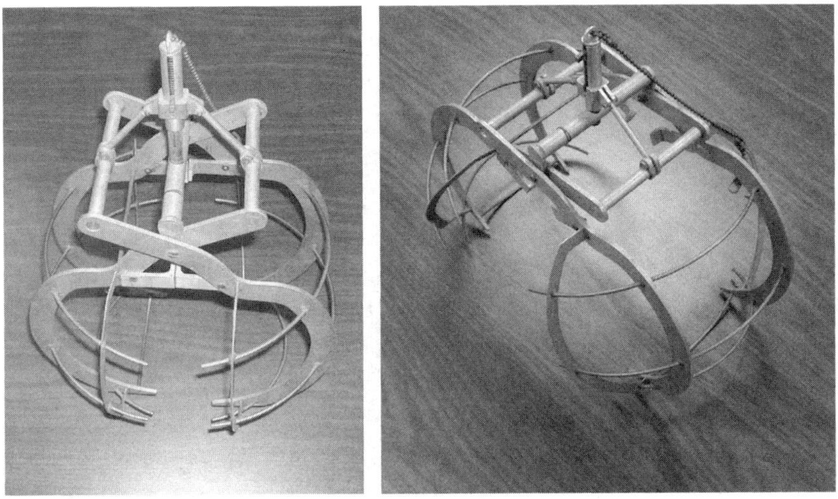

Fig. 18.1 Retrieval claw, closed (left) and open (right).

sonar equipment. It is amazing that the reentry vehicle was located at all. The actual bucket was finally located on October 20, 1971.

The first dive was held on November 5 but was unsuccessful in retrieving the bucket. Heavy seas with winds up to 20 miles per hour and seas 5 to 10 feet high prevented more dives. The ships returned to Pearl Harbor without the bucket on November 15.

Fig. 18.2 Relative size of the claw.

Fig. 18.3 Apache towing support ship White Sands with the Bathyscaph Trieste.

After a number of mechanical problems with sonar and winching, the next dive was on December 1 but further dives were prevented again by bad weather. Finally on April 25, 1972 the bucket was found—it had broken up and some strands of film were streaming out of the wreckage.

During the retrieval, the men in the Bathyscaph Trieste (Figs. 18.4 and 18.5) were able to see that the take-up assembly in the reentry vehicle had partially broken through, thereby revealing the contained film. These men

Fig. 18.4 Trieste II (DSV-1).

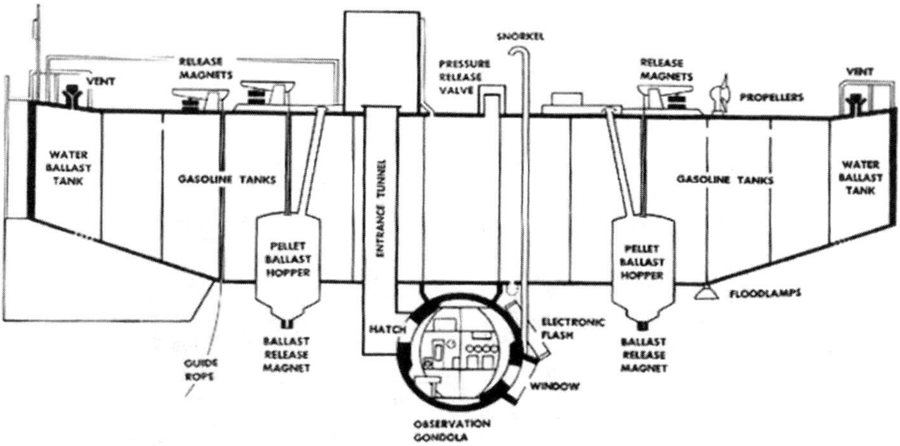

Fig. 18.5 Schematic of the Trieste's internal arrangement.

knew that the purpose of the mission was to retrieve film from a spy satellite. The crew of the support ships did not know this. The CIA had insisted that the reentry vehicle and its contents be hidden from view of the ship's crew as they did not want it known that film was being retrieved. A huge black plastic covering was made ready to cover the remains. Of course, the support ship's crews were quite annoyed at being kept in the dark.

The claw that Len Molaskey designed was put in place around the remaining portions of the bucket visible on the ocean floor. During ascent, short and long pieces of film broke off and sunk. So did portions of the structure of the reentry vehicle and the film take-up assembly. According to the *Quest* article, the claw reached close to the surface after 9 hours from the beginning of the dive.

> By this time, pieces were hanging through the lines of the hook. To minimize surface swell, the hook was lowered to about 35 feet below the surface while a five-person dive team from the White Sands 3000 to 4000 yards away raced to the scene in small boats. The film stacks nevertheless continued breaking apart and disintegrated into a cloud of reddish brown dust about 10 minutes before the divers arrived. All that remained were unusable shreds of film 3 to 6 feet in length hanging from the hook. Nothing suggesting a satellite remained for others to find.[‡]

The Trieste itself was then hauled onto the White Sands and the Apache towed it back to Hawaii (Figs. 18.6 and 18.7). Later strands of film were

[‡]David Waltrop, "An Underwater Ice Station Zebra," *Quest: The History of Spaceflight Quarterly*, Vol. 19, No. 3, 2012.

Fig. 18.6 Trieste being launched from the White Sands.

analyzed by Kodak. They concluded that the film had separated from its support structures on impact with the ocean at approximately 2600 g. The loss was not due to prolonged exposure to seawater, high pressure at the bottom of the ocean, or cold temperatures. Fortunately this was the only occasion during the 15 years of Hexagon missions that it was ever necessary to retrieve a reentry vehicle from the depth of the ocean or anywhere else.

During the 15 years of launches, security remained a paramount concern. Did anyone outside the program know what was going on? Bob Jones, Director of Engineering, remembered a successful launch:

Fig. 18.7 Return of Trieste after a successful dive.

> I was out there one time on business with my wife Stephanie, and we decided to take some vacation time. There was going to be a launch so I said to Stephanie, "Come on up." I told her when we were outside near the gate and near the water, "Just hang around here." I went to the block house. There was a Russian submarine offshore. They were there for every launch, either a trawler or a submarine. The trawler was a fishing boat that had lots of antennas on it. When you checked into a hotel you never wrote down the company you represented because that was secret, as was the launch. When we checked in a few days before the launch, the clerk would say "Launch coming up, huh?"

The last chapter discusses what we all viewed a sad event—the explosion of the last launch.

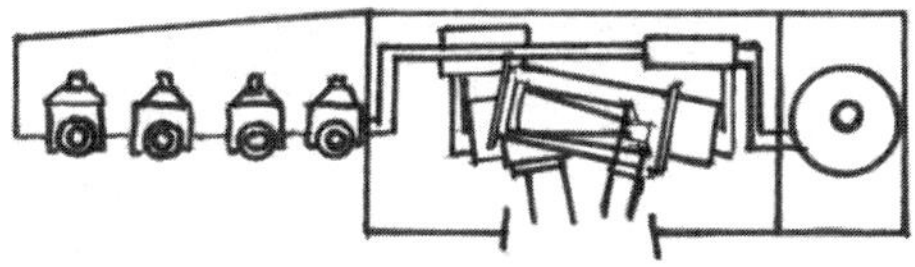

Chapter 19

Explosion

Phil Pressel

A Sad Ending

There were a total of 19 successful flights of Hexagon. Only the 20th and last planned mission failed; it was due to an explosion of the vehicle as it was being launched on April 18, 1986 just a few months after the fatal destruction of the Space Shuttle *Challenger*. This was a bad time for American spaceflight as there were other rocket failures, such as the Atlas rocket carrying a GPS satellite in 1981. A month after the Hexagon explosion a Delta rocket carrying a Goes G weather satellite blew up and years later several Delta rockets failed.

The Hexagon explosion was a terrible mishap. Figure 19.1 shows the explosion approximately 800 feet above the pad. Figures 19.2, 19.3, and 19.4 show damage to adjacent areas. Bob Jones, Director of Engineering said this camera "was probably worth a quarter of a billion dollars." But the event cost more than dollars, as recalled by several Perkin-Elmer employees, including Vic Abraham, the last general manager for the program, and Len Farkas, then deputy director who had worked on the original development of the program and was also instrumental in solving many in-orbit problems. The apparent cause of the explosion is explained in the last section of this chapter.

Vic Abraham's Story

The explosion was an absolutely terrible situation. We (Fig. 19.5) had spent a year before the explosion developing a special bonus program to keep people on the program after the last system vehicle got launched. For example, what would prevent Frank Harrigan (manager of WCFO at Lockheed in Sunnyvale) and his field people or even our key people in Danbury from leaving and taking other jobs instead of waiting for the last system to end? The last system was going to go for two years, believe it or not. We had to keep key people that understood electronic boxes and other things, as well as field people during the whole time of the launch and orbit.

Copyright © 2013 by the American Institute of Aeronautics and Astronautics, Inc., All rights reserved.

Fig. 19.1 Explosion.

Fig. 19.2 Damage to the pad.

Fig. 19.3 Damage to a nearby car.

I and Colonel Phil Datema worked up a beautiful incentive program with both Perkin-Elmer and the government putting money into a pot to give these guys bonuses for the length of time they would stay until after the mission was completed. That was a very rough thing to do because it took money away from profit, but we got approval from Perkin-Elmer. They put this profit money into a special escrow as did the government, the first time they ever did that. It was a unique bonus.

Anyway we were standing there waiting for the blast off. We were in the fallback position and the thing exploded…and that was the end of the bonus plan; we didn't need the field team anymore; we didn't need anybody.

Fig. 19.4 Nearby damage.

Fig. 19.5 Vic Abraham, mid-1980s general manager of the Hexagon program.

It was 8 seconds into launch when it exploded and the wind blew towards the fallback position. I was standing there with a lot of the government people. This cloud of yellow smoke from the boosters was coming our way—we had to run like crazy into our cars and get the hell out of there. I wasn't in the blockhouse this time because I wanted to see it in person, not on camera. I was in the blockhouse for other launches. Then later we didn't know what to do with ourselves. Our hearts went out; what were we going to do with all the people back in Danbury and at Lockheed? It was a terrible situation.

Len Farkas's Story

It was a nice day as usual out there at Vandenberg. The vehicle went off on time. We were about a mile back in the fallback area, not in the blockhouse. There were about 100 people there. It went up…seconds after launch, a big flash…several seconds later, a big boom…then the first words that I heard

Fig. 19.6 Len Farkas, deputy director for the Hexagon program.

over my shoulder were from Harrigan, "Martin Marietta, those sons of a bitches." He apparently had some inkling of a problem on the vehicle in the rocket part of it (a Titan), beyond that I don't know any other details. Then the launch cloud started forming up there; we were told it could be toxic and we'd better start getting out of there. So we all evacuated. I hadn't been to one because I was always in our war room to lend support. It was such a drain going from the excitement to all of a sudden, "Oh my god, what a catastrophe." Fortunately there were no people aboard; then you realize it was the last one (Fig. 19.6).

FRANK'S GRABOWSKI'S STORY

BACKGROUND

Frank Grabowski was an electrical technician on the program for many years. He worked in the Danbury assembly area and often went to perform other tasks at Lockheed in California. His tasks included wiring test equipment for most of the major subassemblies, wiring of electronics in the platen, refurbishing take-up assemblies, and wiring test consoles. He was a witness of the explosion—he had been called upon to help. One of his tasks was to help top-off the nitrogen tanks in our system while it was on the launching pad at Vandenberg. I interviewed Frank because he not only was an eye witness but he was so passionate about his impressions and reaction to the explosion.

Before Frank's trip he had asked our Security office if it was okay if he could videotape the launch from his motel in Pismo Beach located about 3 miles from the launch site. He attended a security indoctrination lecture at Vandenberg, and then was sent up to the 16th level of the gantry to help top-off the nitrogen tanks. He was speechless at first when facing the Titan 34D as it was so awesome. Before he descended he wished it "Godspeed."

When he got back to his motel, he purchased a bottle of champagne and some cups for a toast with his colleagues. There were launch delays, but finally at 10:44 a.m. Pacific time he set up his video camera and started taping outside the motel. Meanwhile his boss at the time, Jules Cohen, and other colleagues from Perkin-Elmer WCFO, including George Manolis, were in the blockhouse near the pad and in contact by a "red phone" with those at the motel.

The Event in Frank's Words

As the launch started I yelled "Go, baby, go!" Then I saw a purple-orange cloud of gas. I realized that there was an explosion. I was devastated. This became the worst day of my life. I threw the camera with the recording into the trunk of my car. I then threw the bottle of champagne into the brush. Later Jules and some others who had been in the blockhouse were escorted away by Security after the cloud of gas had cleared. I went to my motel room and cried. I said to myself, "I don't want to see the videotape." When Jules came back to the motel he insisted on playing it. I was very uncomfortable seeing big chunks coming away from the explosion. What do I do with the tape after I played it? I hid it under my bed.

I got a telephone call in the room from someone saying they were with Security and saying, "I understand you have a video tape of the explosion. Do you have any copies?" I said no. The man said, "Tomorrow at 10:30 go to your hotel in Sunnyvale and bring the tape." I became very nervous. I put the tape in a brown paper bag and went to the hotel as requested. I handed it over to the man and obtained a receipt for it. I still have the receipt. I left the hotel and got my car to go to the airport. I was nervous. I couldn't open the door to the rental car. It was the wrong car. Sweating and sad I flew back to Danbury.

Bob Zarba, one of the vehicle managers asked me if I would go back to Vandenberg with some of the other West Coast staff to help retrieve film and other hardware from the area near the pad. I refused to do it because it was so emotional for me. But the next day I decided to go anyway. It was a sad trip for me.

Several months later in Danbury I was called to go to the main conference room. They were showing the video. They wanted me there as the maker of the video. The video showed the explosion from another vantage point rather than the films taken from the blockhouse. It turned out that watching the video again helped me to get over the trauma since I was not allowed to talk about it at home for security reasons.

To this day I relive the moment every April 18 at 1:45 Eastern time by stopping what I am doing and standing in silence.

Aftermath and Cause of the Explosion

George Manolis recalled that after the explosion, Security had concerns with the possibility that any leftover film could be spread out over a huge area at Vandenberg's Slick 4 launch site (Fig. 19.7). Evidence of film in the area would be a major security problem. A team of from Perkin-Elmer WCFO was assigned to go throughout the base with black plastic bags. "We scoured the area for several days, and picked up all the pieces of film that we could find."

After months of investigation it was found that insulation made of nitrile butadiene rubber that surrounded the propellant to prevent the ultra-hot combustion from reaching the steel rocket case had come loose near a joint between two segments of the rocket. This allowed the hot gases to burn through the case; destruction followed quickly.

The yellow cloud that rapidly approached those watching the launch outside the blockhouse not only was a concern to them but to people outside the Vandenberg base. Liquid fuel was nasty stuff. It could sit in the missile's tanks in a silo for long periods of time, but if it escaped the fumes could be deadly to anybody not wearing protective gear or breathing apparatus.

That big ugly cloud also did some human damage. A letter written to the base commander a week after the explosion captured some of the drama. It was written by a local resident, Martha Boise, who lived nearby downwind of

Fig. 19.7 View showing area of debris.

the explosion. Ms. Boise wrote, "Your rocket explosion last Friday, April 18th has affected my children. They were all exposed to the toxic cloud. They were not given the advantage of any warning or any information on the toxic material in the air surrounding them."

Ms. Boise wrote that her family had called Vandenberg twice about the air quality, and the reply was that the base could not provide any information. She wanted to know why there had been no warning, what the toxic effects of the chemical cloud were, whether there was nuclear material on the rocket, who should they deal with, who should they call in the event of a future explosion, and who would warn them the next time.

Apparently she got her answers, for Ms. Boise wrote the base commander a follow-up letter a few weeks later after a visit by some Air Force officers to reassure her that the military was taking the situation seriously and that the base had changed procedures about warning local residents in case of emergency.

Even those of us who were not at the launch pad to witness this awful event felt the loss personally. I found out about the explosion a few minutes after it occurred. I was walking in the hallway of our building in Connecticut when Paul Petty, the general manager of the division at the time, saw me and told me of the failure. I could not believe it, but Paul confirmed it. I immediately choked up and tears came to my eyes. I rushed to my office where, in solitude, I cried. I had devoted many years to this. It was like a family member dying. It would be the end of an era for all of us.

A Happier Memory

The final launch was painful for all us working on Hexagon. Allow me to conclude the story of how Hexagon came to be with a pleasant memory. The *first* launch was on the morning of June 15, 1971 from Vandenberg Air Force Base in California. Some of us gathered in our war room in Connecticut a short while before the scheduled launch. I sat on a side table near one of the few audio speakers over which the launch sequence would be broadcast by the Air Force, who was in charge of the launch. Along came our chief, Mike Maguire, and he sat next to me at the table.

As it turned out, the voices coming through the speakers were infrequent and nondescriptive. All that came through were phrases like "one minute to launch," then "rockets have fired," then "liftoff has occurred," and a slight amount of technical data, finally followed by "system is in orbit." But for everyone at Perkin-Elmer, this was a very exciting event. We had been working and planning for this day for six years. Mike went around the room and shook everyone's hands. That is how I like to remember our experience on Hexagon.

Appendix A

HEXAGON KH- 9 LAUNCH DATA HISTORY

PHIL PRESSEL

Boosters: Titan III-D 1971 to 1984, then the Titan 34-D.

Date	Lifetime (days)
June 15, 1971	52
Jan. 20, 1972	40
July 7, 1972	68
Oct. 10, 1972	90
Mar. 9, 1973	71
July 13, 1973	91
Nov 10, 1973	123
April 10, 1974	109
Oct. 29, 1974	141
June 8, 1975	150
Dec. 4, 1975	119
July 8, 1976	158
June 27, 1977	179
Mar. 16, 1978	179
Mar. 16, 1979	190
June 18, 1980	261
May 11, 1982	208
June 20, 1983	275
June 25, 1984	115
April 18, 1986	0 (Exploded after takeoff.)

Copyright © 2013 by the American Institute of Aeronautics and Astronautics, Inc., All rights reserved.

Appendix B

The Twister

Phil Pressel

The twister was the invention of Don Cowles (Fig. B.1), one of my mechanical engineering colleagues at Perkin-Elmer. Without it and the air bars, it is unlikely that this program would have succeeded. Certainly technical operation and configuration of the program would not have been possible without the twister. It was the key to permit the film to not only rotate with the rotating optical bar, but also to travel linearly past the focal plane. Unfortunately personal information is unobtainable as Don died years before this book was begun.

Don was a long-term mechanical engineer at Perkin-Elmer. He was on the program from its inception and study phase though the mid-1970s. Don was also involved in the design and testing of mechanisms in other major subassemblies of the system. His inputs would have made for interesting and enlightening reading. But we have his patent, which describes how the twister worked (Fig. B.2).

Copyright © 2013 by the American Institute of Aeronautics and Astronautics, Inc., All rights reserved.

Fig. B.1 Don Cowles, mechanical engineer.

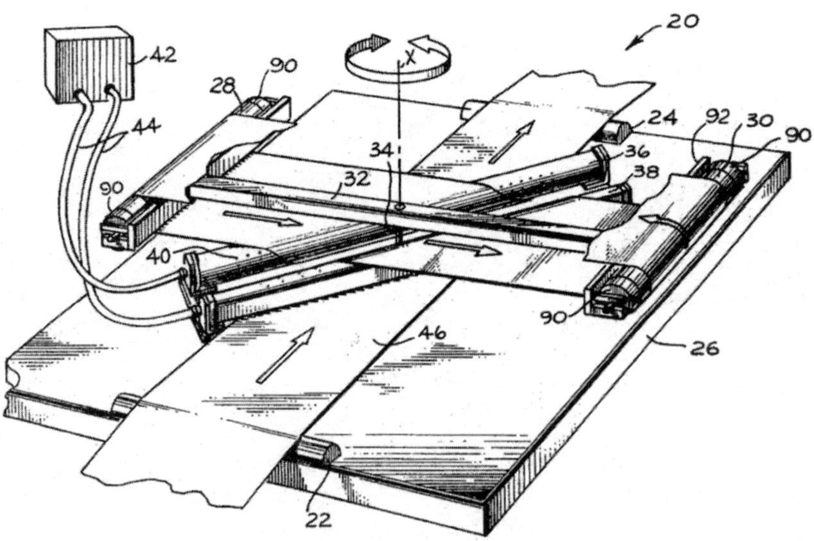

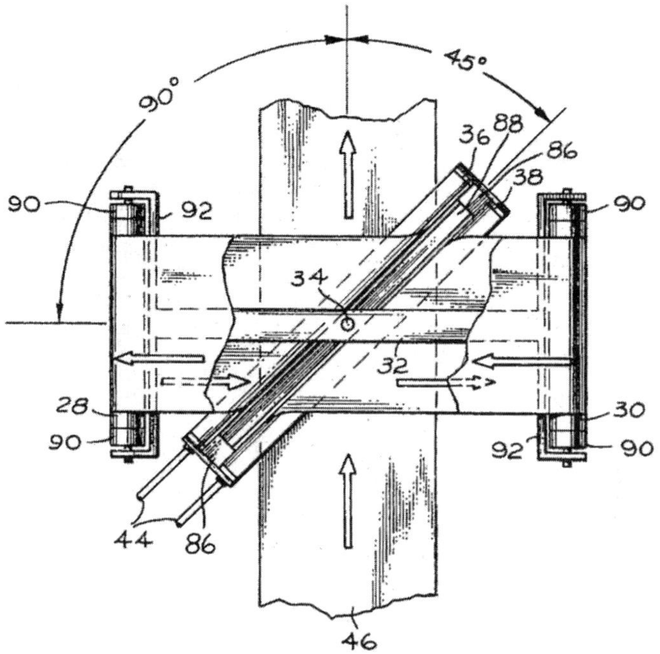
90°
45°
36
88
86
38
90
92
90
34
32
28
30
90
92
90
44
86
46

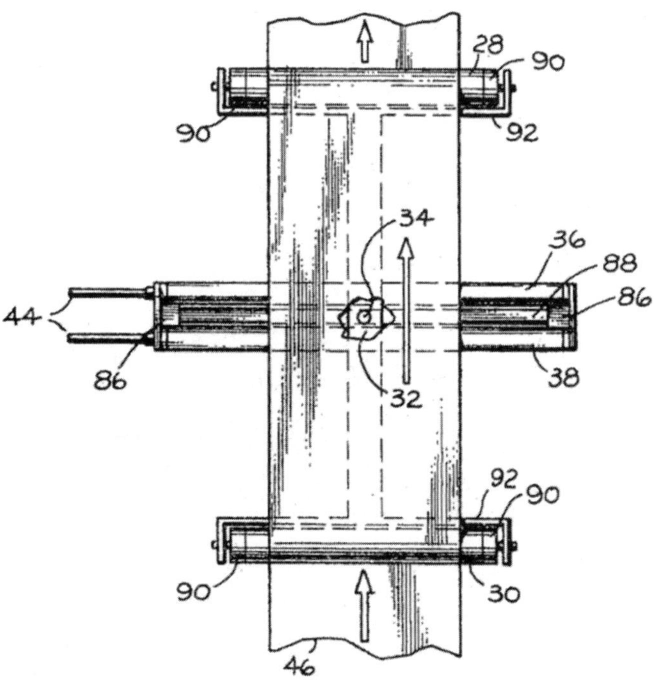
28
90
90
92
34
36
88
86
44
86
32
38
92
90
90
30
46

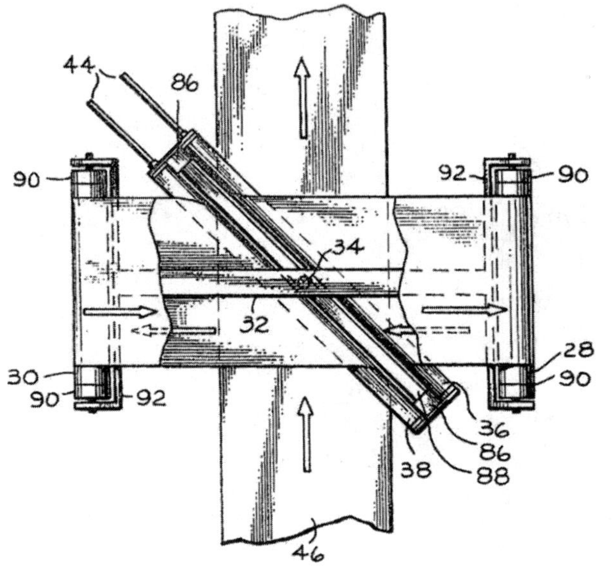
44
86
90
92
90
34
32
28
30
90
90
92
36
86
38
88
46

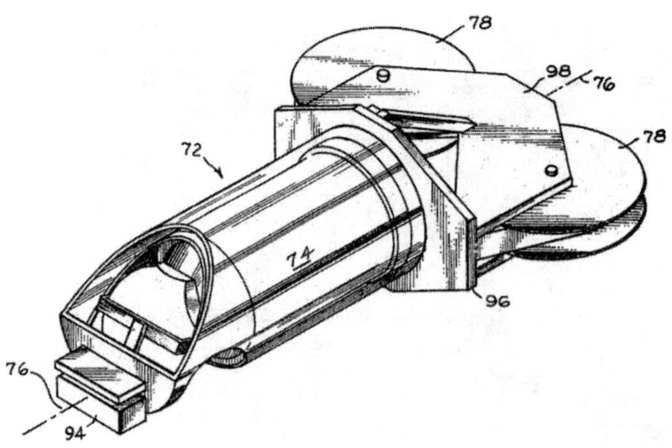
78
98
76
78
72
74
96
76
94

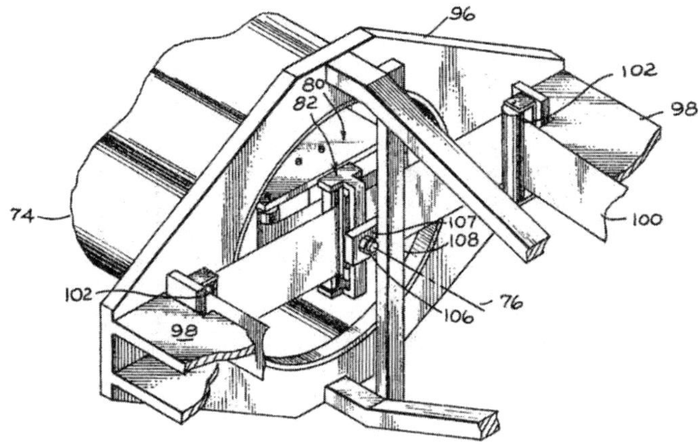
96
102
80
82
98
74
107
108
100
102
98
106
76

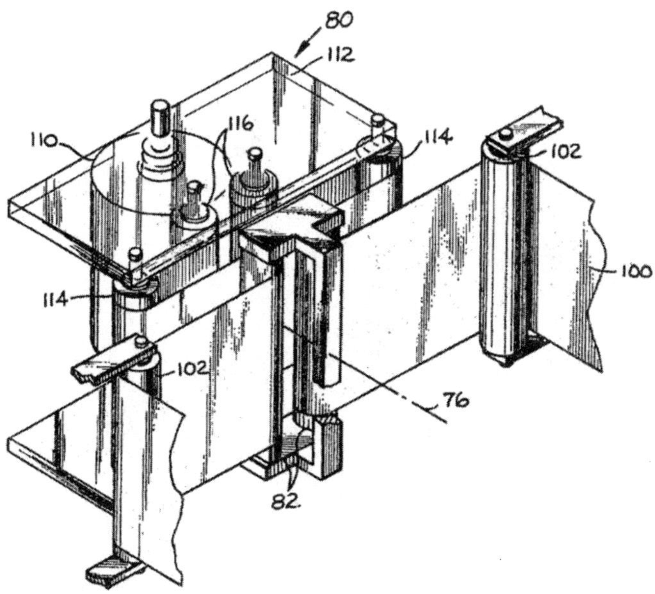
80
112
110
116
114
102
100
114
102
76
82

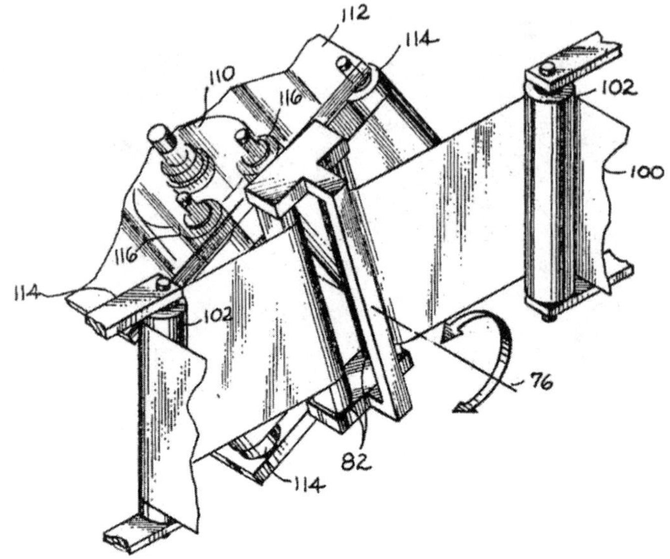
112
114
110
116
102
100
116
114
102
76
82
114

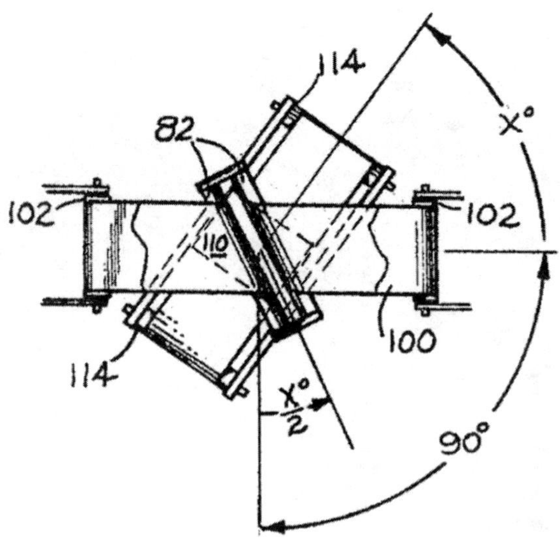
114
X°
82
102
102
110
100
114
X°/2
90°

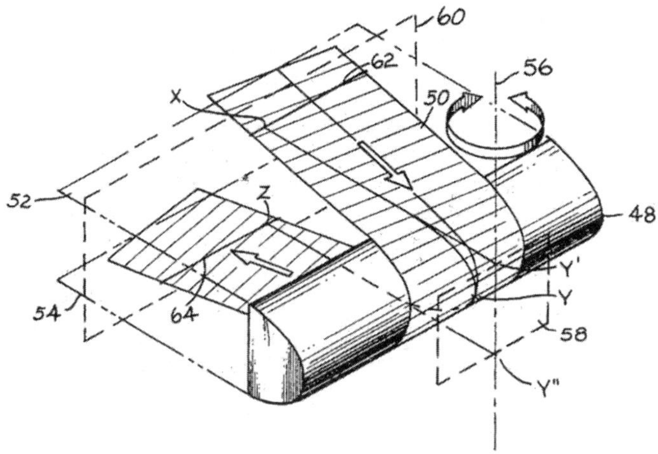
60
62
56
X
50
52
Z
48
Y'
Y
54
64
58
Y"

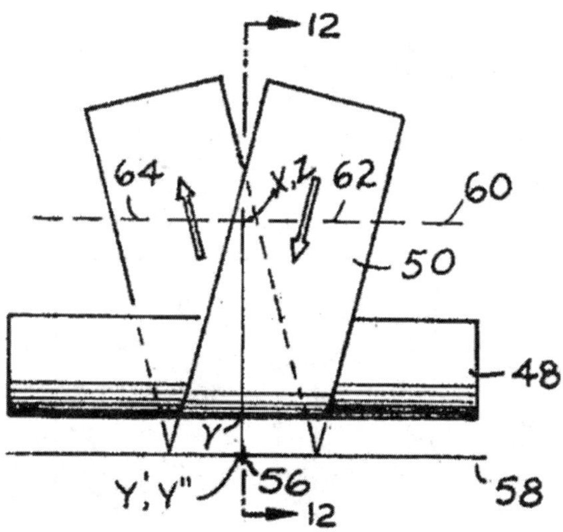
12
64
X,Z
62
60
50
48
Y
Y',Y"
56
12
58

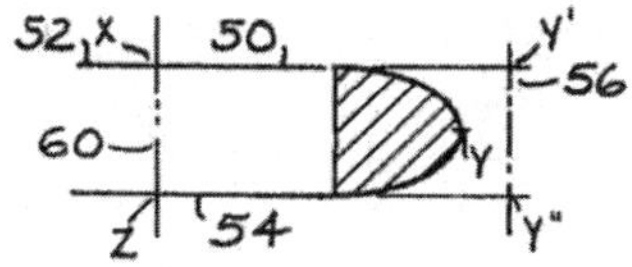
52,X
50,
Y'
56
60
Y
Z
54
Y"

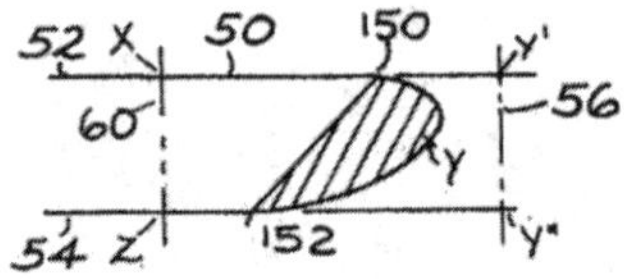
52
X
50
150
Y'
56
60
Y
54
Z
152
Y"

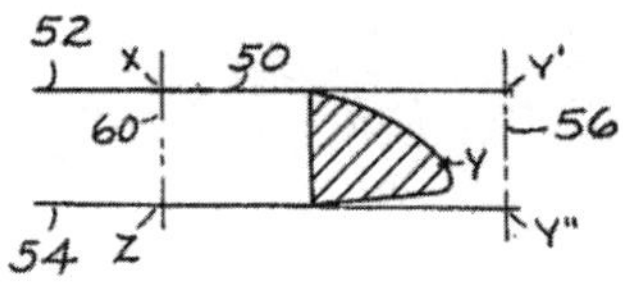
52
X
50
Y'
60
56
Y
54
Z
Y"

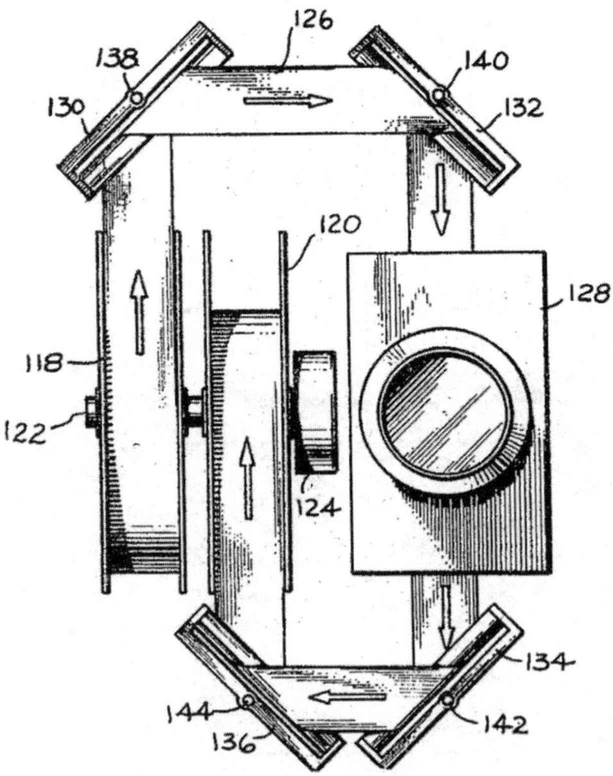
126
138
140
130
132
120
128
118
122
124
134
144
142
136

Appendix C

Servos, Image Motion Compensation, and Precision Velocity Control

Marty Yellin, Dick Carricato, and Richard Labinger

Servos

This discussion of servos includes details about the drive capstans, fine tension sensors, metering capstans, and take-up servos.

Photographic exposure occurred as the film was transported through the platen that oscillated in synchronization with the optical bar during the exposure time. The film was driven on and off the platen by two velocity servo-controlled drive capstans. These are brute force devices that drive the film at the prescribed velocity profile, although not to the accuracy required during exposure. Between the drive capstan and the metering capstan in the platen, there was a device called the fine tension sensor—very similar to the coarse tension sensor (see Fig. 7.6).

It was necessary that the drive capstan velocity be exactly the same as the metering capstan velocity and that the metering capstan velocity be precisely synchronized to the image motion. If the drive capstan velocity were different from the metering capstan velocity, then the film would lose its correct tension and focal plane position. The fine tension sensor is at null when both the drive and metering capstans are at equal velocity. If the drive capstan is going faster, the fine tension sensor would displace towards ground applying a negative signal to the drive capstan to slow it down. Conversely if the drive capstan is going slower, the fine tension sensor will displace away from ground and send a positive signal to the drive capstan to speed it up. The drive capstan incorporated an electrical integrator in its forward loop so that it had zero velocity error for a fixed error command.

The metering capstan's main function was to very accurately control the film velocity, in accordance with the instantaneous velocity of the image on the ground. This function is called *image motion compensation* (IMC). The tracks on the optical bar encoder, in addition to the incremental tachometer track and

Copyright © 2013 by the American Institute of Aeronautics and Astronautics, Inc., All rights reserved.

the motor commutation tracks, contained the trigonometric functions of the optical bar angle. These pulse rate signals were demodulated by frequency to voltage converters to derive analog signals used in the computations of the IMC equations. The IMC profile generated on the optical bar encoder was fed to the metering capstan servo, which was then position or "phase locked" to the optical bar. Even though we didn't invent optical encoders, we applied them in a manner that was deemed revolutionary at the time.

In essence, the approach used two digital counters. One was fed by the optical bar encoder track. The other was fed by the metering capstan encoder track. The digital difference represented the position error between the optical bar or ground image and the metering capstan film position. This difference, or position error signal, was fed to the metering capstan to lock the film speed and position exactly with the image speed and position.

The limited rotation platen servo presented one of the greatest challenges of all. During photography it had to rotate exactly in synchronization with the optical bar image rotation. Because it also was coupled to the metering capstan containing the film, the platen could not continue to rotate like the optical bar did during the nonimaging period because the film would get twisted. Instead, at the end of the imaging rotation angle, the platen had to reverse direction and get ready for the next photography cycle at the exact time and position that the optical bar started its imaging sequence. In order to accomplish these two very different modes, the platen contained two different feedback position indicators, a capacitive nulling sensor used during imaging, and a rotational position encoder used during nonimaging. The air bars and twister were key mechanical elements that permitted the platen's reversal of direction.

During photography the platen was configured as a very accurate position servo synchronized to the optical bar rotation position plus and minus an additional IMC factor. In order to achieve that goal, we invented a capacitive nulling sensor consisting of two conductively coated glass plates as shown in Fig. C.1.

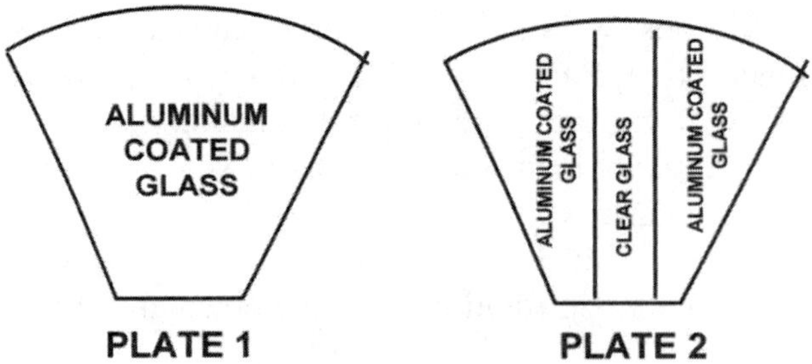

Fig. C.1 Capacitive sensor plates.

Plate 1, which was a continuous metal-coated glass plate, was locked to the reference optical bar rotation angle that essentially represented image rotation. Plate 2, which was a split pair of metal coatings on a glass plate, was mounted to the rotating platen. It acted as a nulling detector, in that equal and opposite voltages were fed to each half of the split plates. If, for example, plate 1 was grounded, then any displacement between the center of plate 2 and plate 1 would cause an unbalance capacitive current signal, which represented the error signal used to drive the platen motors until the error was reduced to zero. When this occurred, the platen position exactly tracked the rotating image generated by the optical bar. In actuality, however, we did not want the two servos (platen and optical bar) to track exactly but instead to differ by the required rotational IMC. To accomplish this, we replaced the ground on plate 1 with a sinusoidal signal that represented the inverse of the IMC signal. This effectively caused the platen servo to null out with zero positional error at the exact image rotation position. During nonimaging, the noncontacting glass plates were not used. Instead, the encoder on the platen rotational axis would be used to close the loop during recycle via commands derived from the optical bar encoder. This would bring the platen to the right position to start photography again.

The film take-up servos were similar to the supply servo but because there were four take-ups for each film supply, the change in velocity and inertia was much less than that of the supply servo. Again there was a coarse tension sensor to measure the difference in velocity between the take-up spool and the looper film exit velocity. As discussed for the supply, any difference in speed caused a displacement of the coarse tension sensor that in turn provided a corrective bias signal to the take-up servos to control the motor angular velocity to make the take-up film velocity match the looper exit film velocity. Unlike the supply, the take-up required a builder roller to assure a proper, nontelescoping stack of film. This provided a convenient mechanism in which to incorporate a take-up radius sensor for use in the take-up servo system.

IMAGE MOTION COMPENSATION

IN-TRACK IMC

In-track IMC means normal to the scan direction across the film. The vehicle moved over the ground at more than 4 miles per second. This translated to about 3 inches per second of image motion moving across the film at the focal plane. The direction is at a right angle to the direction of motion caused by the scanning optical bar, and its magnitude reduces as the scan angle moves away from nadir. This image motion can be compensated

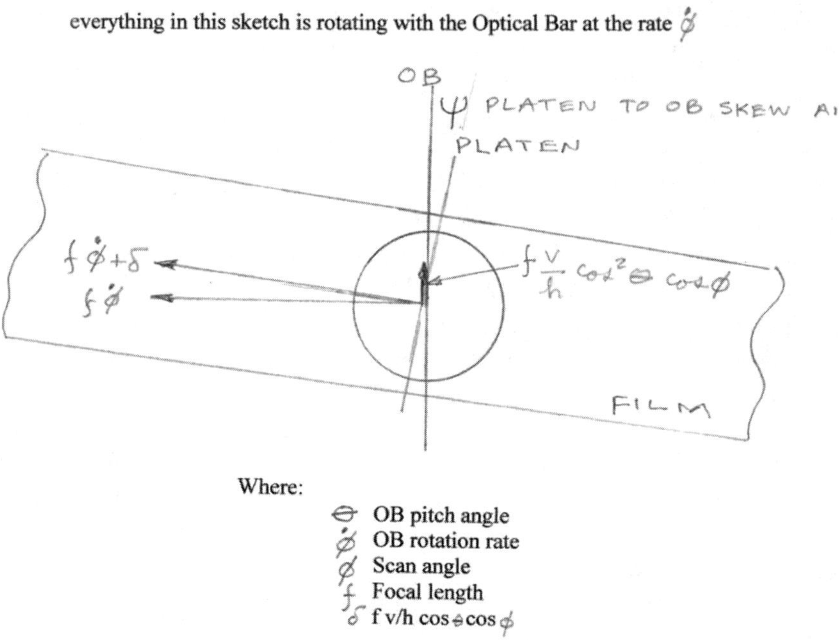

Fig. C.2 Cross film and along film image velocity vectors.

by rotationally skewing the platen with respect to the optical bar. At nadir the required skew angle pound per square inch is about 0.09 degrees, and this reduces to half that value at +/−60 degrees from nadir.

All of the in-track errors are compensated by the platen to optical bar skew angle. This angle, which is a function of the optical bar angle and *V/h*, is computed by the camera system, and then used to precisely control the platen servo system. The skew angle is shown schematically in the Fig. C.2. Angle theta is the 10-degree stereo angle. The skew angle is controlled by a capacitive sensor between the platen and the optical bar. The angle changes the relative rotational position of the platen and the optical bar about the axis of rotation. After applying platen skew angle ψ, film speed must be increased to match the image. Everything pictured in Fig. C.2 is rotating with the optical bar at the rate ϕ. The letters OB and Pl refer to their respective centerlines of rotation.

Cross-Track IMC

Cross-track IMC refers to the along-the-scan direction, along the film. All of the image motions in the cross-track direction were compensated for by the metering capstan servo. This high-precision servo system was located on the rotating platen and was as close as possible to the focal plane. All of the image motion contributors for each camera were computed and then added to the velocity command input to the servo. The integral of the sum of these terms is added to the phase difference input of the phase locked loop.

Each camera, one looking forward by 10 degrees and one looking backward by 10 degrees (therefore the stereo angle was 20 degrees), had its own set of image motion terms that had to be compensated for. The terms were as follows:

The constant velocity term. Two components of the system whose errors are corrected by a single resistor value are the metering capstan diameter and the effective focal length. The resistor is called the *knob*. One of the early decisions on the program was that there would be no electrical adjustments. These are usually thought of as potentiometers with knobs. So this adjustment, which is accomplished by a fixed resistor value, is called the knob.

The term that results from Earth rotation, V_y/h. A reconnaissance satellite usually travels in a north-to-south orbit allowing Earth to rotate below it. This gives coverage of almost the entire surface of Earth. Orbits close to 96 degrees precess at about the same rate as the Earth about the sun. This keeps the sun in the same position for taking pictures each day while on-orbit. Earth rotation results in a maximum effect of about 0.12 inches a second. The terms that arise from the coordinate transformation of points on Earth to points in the focal plane. Only those terms that are significant, correctable, and not already compensated are included. These terms, which are zero at nadir, can reach a maximum of about 0.3 inches a second at 45 degrees.

The terms to correct the speed of the film for the platen skew angle. Remember, the platen is offset slightly from the optical bar to align the film velocity vector with the image velocity vector. Therefore, the film speed must be increased slightly, about 0.4 inches a second, to match the image speed.

Noncorrected or noncorrectable terms:

The rate of change of altitude while exposing the film.

The platen skew rate. Because the platen rotates the film with respect to the image by about 0.4 degrees during photography, there is a component of film velocity in opposite directions at the top and bottom of the slit. This velocity is zero at nadir, but increases to about 0.1 inches per second for a 45-degree scan angle at the edge of the film.

Images at the edge of the film in one camera show up at the center of the film in the other camera, where the performance is optimum. It has never been determined whether the stereo angle of 20 degrees that results in a 3.5-inch frame separation between the two cameras was fortuitous or intentional. The photo interpreters probably gave a range with bigger being better.

PRECISION VELOCITY CONTROL

Figure C.3 shows a schematic of the looper film path. The eight rollers on the carriage move with the carriage. The other rollers are fixed to the looper structure.

Film with a speed V_{CFP} (CFP stands for coarse film path) from the supply enters the looper at the top left of Fig. C.3. Film with a speed of V_{CFP} exits to the take-ups from the looper shown at the lower left. If the capstan on the right maintains the film speed at V_{CFP} from top to bottom, the looper carriage will remain stationary. If the speed of the capstan is greater than V_{CFP} the looper carriage moves upward. It moves downward when the capstan speed is less than V_{CFP}.

The high film velocity at the capstan is labeled FFP (fine film path) in Fig. C.3 and takes place in the platen (at the focal plane). The high-speed film enters the platen as the looper moves upward thus using film that the supply had fed into the looper. After film exposure, the film speed reduces and the film is fed back to the other side of the looper carriage to go into the take-ups.

The metering capstan was the servo controller that set the speed of the film as it passed by the camera's slit and exposed. All we had to do is maintain the accuracy to one one-hundredth of one percent.

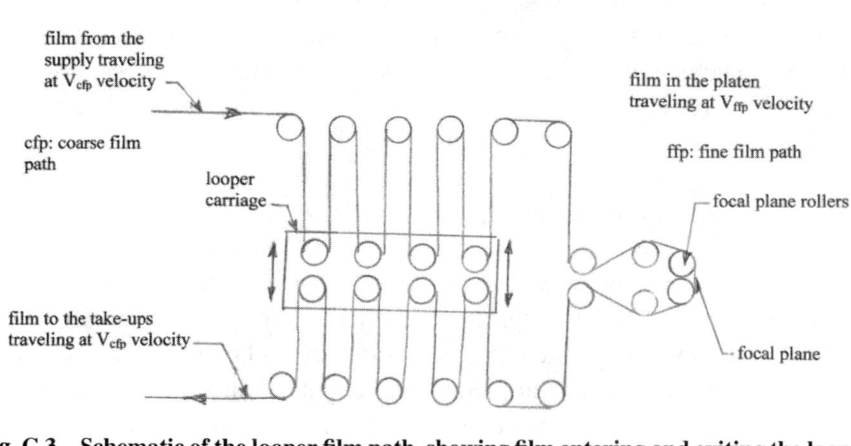

Fig. C.3 Schematic of the looper film path, showing film entering and exiting the looper.

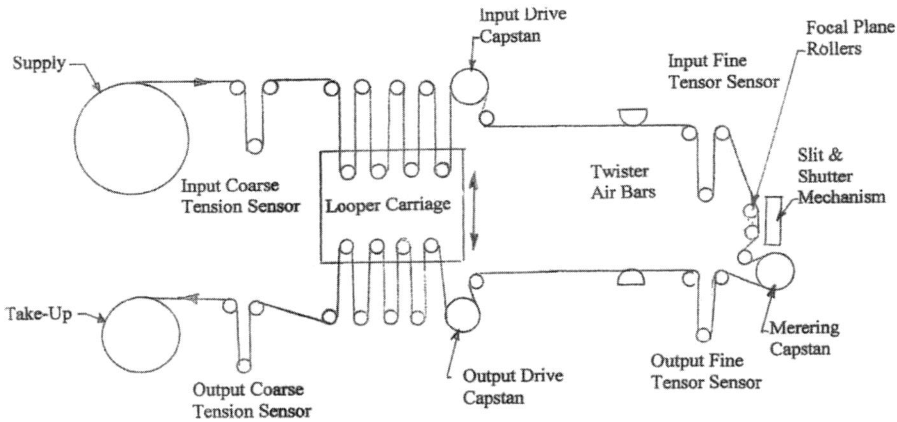

Fig. C.4 Schematic of entire film path.

Bob Landsman (Figure 2.7), our engineering manager anticipating our skepticism, said that when we wish to achieve zero velocity error in a servo it is customary to close a position loop around it. If an incremental encoder on the optical bar (representing image displacement) were phase locked to an incremental encoder on the metering capstan (representing film displacement) this would assure perfect velocity lock between the film and the image. This statement would be true if many other things in the system were perfect. Of course they were not.

The metering capstan motor was designed as small as possible to isolate the servo from the various motor-generated disturbances. The metering capstan was isolated from film disturbances by two film tension isolators located between the metering capstan and the twister. To further reduce the size of the metering capstan motor, two drive capstan servos were placed at the fine film path ends of the looper. The two tension sensors, which doubled as the tension isolators for the metering capstan, provided position feedback to the two drive capstan servos, thus driving their velocity errors to zero.

Additional tension sensors were added between the supply and the looper and between the take-up and the looper. These tension sensors also provided the film position signal to the supply and take-ups to drive their velocity errors to zero. These tension sensors, of course, also maintained the tension constant throughout the entire film path. With these additions the entire film path can be represented schematically as shown in Figure C.4.

Appendix D

Thermal Control, Modeling, and Analysis

Karl Reseck*

As the preliminary design phase progressed, an effective group of camera math models was established, basically as follows:

Coarse models were used for supply assembly and typical take-up assembly.

A coarse model of the TCA and the supply assembly was developed by Perkin-Elmer for use by Lockheed. Lockheed incorporated this model into their integrated spacecraft model. Perkin-Elmer also used this coarse model for parametric studies, such as estimating the approximate effect of changing a specific coating.

A detailed model of the TCA was combined with the coarse model of the supply assembly. This was the primary model used by Perkin-Elmer for thermal design and mission temperature predictions. Thermal boundary conditions used by Perkin-Elmer were obtained from results of Lockheed's analyses using their integrated spacecraft model.

Detailed models were created for optical items, including primary mirror, corrector plate, and field group lens assembly. Boundary conditions for thermal analyses using those detailed models were obtained from results of analyses using the detailed model of the TCA.

Implementation of Thermal Design Requirements

Throughout the camera thermal design phase, Lockheed and Perkin-Elmer performed coordinated thermal analyses applicable to the TCA and supply

Copyright © 2013 by the American Institute of Aeronautics and Astronautics, Inc., All rights reserved.
* Karl was the manager of Perkin Elmer's thermal analysis section.

assembly. Results were exchanged, analysis inputs were coordinated, and changes in thermal properties were selected for subsequent analyses.

Design of the baffles outside the viewports of the two optical bars required repeated iterative analyses by both Lockheed and Perkin-Elmer. Many combinations of orbital, environmental, and operating profiles were considered for various candidate baffle geometries.

Thermal analysis cases included various camera orbital operation modes. One mode was cameral nonoperating (low-power dissipation standby mode), low level of Earth albedo, low level of Earth infrared radiation, orbit selected to produce minimum amount of direct solar illumination of camera viewports, corrector plates for both cameras facing in the opposite direction of the Earth (that is, facing the interior of the vehicle). Another mode was to have the camera operating at maximum optical scan angle (high camera power dissipation mode), high level of Earth albedo, high level of Earth infrared radiation, orbit selected to produce maximum amount of direct solar illumination on camera viewports, and both optical bars rotating as constant rate. As the optical bars rotated, the corrector plates viewed/scanned the entire viewable spacecraft environment, including thermal baffles outside the cameras; they also viewed the Earth albedo, Earth infrared radiation, and occasional direct solar radiation.

Thermally induced distortion of various optical components and effects on optical performance were estimated for several conditions. The analysis sequence was as follows:

For camera thermal environment, use results (temperatures and some heat fluxes of Lockheed's analyses from their integrated spacecraft model).

Impose those Lockheed results as boundary conditions in Perkin-Elmer's analysis using its detailed model of the TCA.

The results of Perkin-Elmer's analysis were available to use as boundary conditions in detailed model analyses of selected optical items.

Perkin-Elmer's analyses using the detailed models produced detailed temperature distributions within the individual optical elements.

These temperature distributions were analyzed by the structural analysis group to predict the corresponding thermal/mechanical distortion of the optical elements.

These distortion profiles were used by the optical analysis group to estimate optical performance degradation parameters, including modulation transfer function (or MTF, a set of numerical measurements that designate the resolution of the images) and defocus.

For the take-up assemblies, Perkin-Elmer supplied a thermal model and power dissipation profiles to Lockheed. Lockheed incorporated this model into its appropriate system-level model and performed all take-up thermal analyses for orbital operations. Accordingly, Lockheed assumed responsibility for achieving the required 70+/−23°F requirement.

As the program progressed, a formal interface control document between Lockheed and Perkin-Elmer evolved. All camera thermal design parameters such as coatings, materials, and insulation blankets were selected.

Primary Mirror

The primary mirror for Block I (flights 1–6) cameras was manufactured from high-purity fused silica glass made by the Heraeus Company in Germany. These mirrors had very uniform thermal properties, including expansion and conductivity. Optical performance at spatially uniform temperatures was excellent. The disadvantages of these mirrors were relatively high thermal expansion, high weight, and high cost.

For Block II (flights 7–12) it was decided to change the material to ultra low expansion (ULE) glass made by Corning. Technical and pricing information indicated this change would offer significant program benefit. Mirrors were manufactured and installed in the first optical bar of flight 7.

Optical performance at room temperature was consistent with prior optical bars. But unfortunately performance at +/−23°F was terrible, indicating large defocus and some tilting. A task force was established to identify the problem.

Engineers from Perkin-Elmer visited the Corning facility and observed the process of manufacturing a mirror. It was formed by a slow process of accumulating sprayed, molten ULE material onto a rotating table. Multiple nozzles were employed and many revolutions of the table occurred before a mirror's thickness was achieved. It was clear that mirror properties could vary, especially in the radial direction. Corning supplied measured coefficient of thermal expansion data obtained from samples of a prior mirror; radial variation was demonstrated. Simple mirror-bending calculations using these data showed that substantial mirror focus change would occur over a 3°F change. The ULE was very likely the cause of the optical bar thermal test failure.

Next, engineers from Perkin-Elmer visited the Heraeus facility in Germany. Quality control and the process of measuring spatial distribution of material properties were regarded to be excellent. The decision was then made to resume use of Heraeus mirrors in all optical bars for Hexagon.

CONTROL OF GAS FLOW RATE

Control of the gas flow rate from the supply assembly to the film path is also important in thermal design. During camera nonoperating periods, the gas pressure in the supply assembly could be significantly higher than the pressure in the downstream film path. On/off gas flow into the film path is controlled by a door/clamp at the exit of the supply.

During camera operation, the door/clamp is opened permitting both film and gas to flow into the film path. An early program question was would the rate at which water vapor flows out of the supply assembly during camera operation be so high as to cause drying of the film on the film rolls?

To answer this question, fluid flow analyses were performed for various depths (through the thickness of film layer) of the film passage (across the 6.6-inch wide film). Throughout the mission, partial pressure of water vapor in the supply assembly was assumed to be the partial pressure of water at the relative humidity of the water in the film. Maximum total nitrogen pressure at the beginning of the mission was taken to be 15 pounds per square inch. At the end of the mission, pressure in the film path was assumed to be zero (vacuum) for all analyses.

THERMAL CONTROL COATINGS

A principal objective in the camera mechanical design was to achieve spatially uniform temperatures in order to minimize thermal distortions and misalignments. This was to be done while the camera was exposed to a thermal environment of large temperature variations, in both location and time. Camera structural members needed to be thermally decoupled from this nonconstant environment. Radiation decoupling was achieved by using a low-emissivity coating on the exterior of the structural member; if this was not sufficient, then applying insulation might be required. To supplement the exterior coating the interior of the structural member was coated black, which increased radiation coupling within the member. Additionally, thermal conduction within the structural material reduced temperature differences in the structural member; employing a structural material with high thermal conductivity also enhanced this process.

These criteria were implemented, for example, in the optical bar structural tube, in the structural frame for the TCA, and in the structural enclosure for the supply assembly. The structural material was aluminum, which met the objective of high thermal conductivity and low emissivity on exterior surfaces. Low emissivity was achieved by polishing surfaces to be shiny and by visually comparing it to a sample test coupon that had its emissivity measured by an instrument called an emissometer. The interior surfaces of the optical bar and supply were painted black; this coating had the additional advantage of reducing reflections of unwanted light inside these items. No thermal control coating was applied to the interior surfaces of the TCA's support frame as it was totally enclosed.

Appendix E

S-CUBED

WALTER FOWSKI, VIC ABRAHAM, AND PHIL PRESSEL

Originally, an Itek mapping camera was used to determine where photographs of Earth had been taken. On the 17th flight system a new pointing camera was installed so that photo interpreters would know more precisely where the Hexagon optical system was looking during operation. A star sensor or spaceborne stellar sensor known as "S-cubed" measured Hexagon's orientation with respect to the celestial equatorial reference frame. This sensor recorded the observed pattern of stars as faint as ninth visual magnitude and compared this pattern to all patterns that possibly could be observed from the spacecraft's orbit as stored in the onboard computer. It was first used in the 1982 time frame.

When a match of stars was found, the pointing of the Hexagon system's line of sight could be determined using complex computer algorithms. The pointing accuracy was on the order of one arc second, with a stability of subarc seconds.

The approximate size of the sensor was 20 inches long and 8 inches in diameter. Two of these sensors were mounted on the forward corners of the Hexagon support frame. The mechanical arrangement of the sensor is shown in Fig. D.1.

Larry Cassidy was program manager for S-cubed. Paul Yoder (Fig. D.2) was an optical engineer who not only was involved in its design but actually taught many of us in our early days at Perkin-Elmer about optics. Paul has written many books on optics and on basic techniques to analyze and mount all kinds of optical elements such as lenses and mirrors.

S-CUBED ELECTRONICS

The S-cubed system was not flown until the last four systems, starting with system 17. Each system had two S-cubed cameras, each having its own S^3 DPE and a single S^3 TPC. The use of the S-cubed camera coincided with the

Copyright © 2013 by the American Institute of Aeronautics and Astronautics, Inc., All rights reserved.

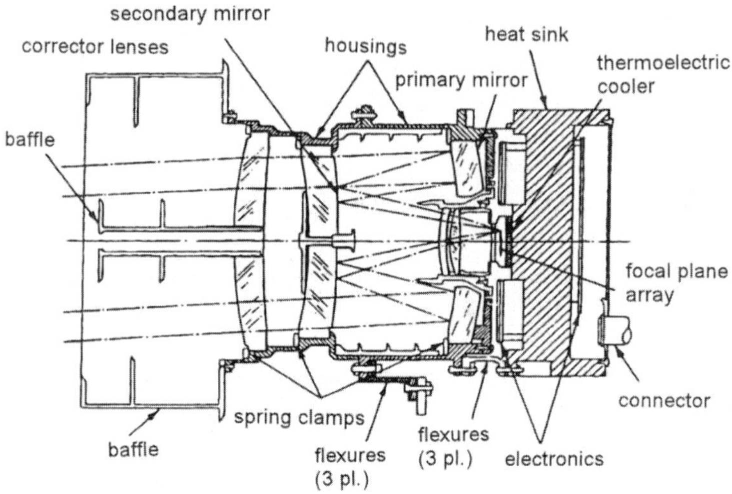

Fig. D 1 Mechanical arrangement of the S-cubed camera.

use of an increased size looper (called the super looper) and an increase in film capacity to 60 miles (30 miles per camera). The following electronic boxes were associated with the S-cubed camera:

- S-cubed data processing electronics mounted on the mainframe provided processing, storage, and transmission of S-cubed camera star data
- S-cubed thermal and power control box provided the command interface, power to the S-cubed focal plane electronics and controlled the coolers in both S-cubed cameras

Fig. D 2 Paul Yoder, optical engineer.

Stereoscopic imagery of Earth is great if you know where you are looking. Hexagon employed a pair of stellar index (SI) cameras made by Itek for the first sixteen missions. The cameras would periodically take pictures of the heavens, and these pictures would be compared to known star maps to determine the exact pointing of SI. Knowing the angular relationship between the SIs and the main cameras meant that the location of any object photographed could be readily determined. During the intervals between SI fixes, the gyro data would be used to correct for spacecraft motion relative to the ground. In the mid 1970s we embarked on the design of the first solid-state star mapper to be space-qualified.

Solid-state Star Mapper

This mapper design project was identified as the S-cubed program. This program involved mounting two cameras on the main Hexagon frame that held the optical bars so that interlock errors between the main cameras and the S-cubed cameras were minimized. The cameras consisted of a stray light baffle, optics, focal plane with detectors and supporting electronics, and a heatsink/structure assembly.

At that time, the charged coupled device (CCD) area arrays that are now in cameras and cell phones were still a dream. What did exist were linear arrays consisting of either photodiodes or CCDs and small photodiode area arrays. John Shamaly and his team of detector experts evaluated our options and, after selecting the Fairchild 1728 CCD array, designed the low-noise focal plane electronics that needed to be mounted very close to the detectors. The focal plane consisted of six linear arrays, each containing 1728 CCD pixels, precisely mounted on a substrate and mapped for exact pixel location in a custom-designed hermetically sealed package with an optical window. The array was cooled by a two-stage thermoelectric cooler and mounted to a large beryllium heat sink. The annulus-shaped focal plane electronics surrounded the array but was thermally isolated.

Operation of the S-cubed system was based upon stellar motion roughly perpendicular to the axis of the linear CCDs resulting in multiple crossings. Each crossing resulted in a number of hits on a few pixels of each array from which the star centroid could be determined. This data was then transmitted to the ground and compared with star maps to define the actual pointing of the main cameras at any given time. Calibration maps of the focal plane were used to reduce residual errors to approximately 1 to 5 arc second. The program was extremely challenging, successful and eventually led to our company entering the solid-state star tracker business in the late 1980s.

VIC ABRAHAM'S STORY

SELLING THE SYSTEM

After flight 20, the last planned Hexagon flight, we asked the CIA, "How are you going to do the next reconnaissance mission?" Mr. X said "We can do this with the new system just as well." The new system was a real-time reconnaissance camera that was to replace Hexagon, the last film-based camera. The Air Force, and a lot of guys from universities who were cleared, said "No, the next system can't do the mapping." They had to order another one of our system or at least put together the spare for an extra system.

For 2 years we tried to sell that. I remember going to Senator Goldwater when he was charge of the Senate Select Committee on Intelligence, and he had a bunch of staffers that worked for him and we had a guy on our board of directors, Dr. Townes from Cal Tech who was the guy that invented the laser. He was a good friend of Weinberger, the Secretary of Defense at that time. I had to brief Townes, and then Townes got me a meeting with Weinberger. Of course this was in conjunction with the Air Force because they were behind us all the time. Anyway [name redacted] thought I really overstepped my bounds by going through political channels to try to keep our system alive and trying to get another block [a series of flights] contract. He tried to pull my tickets [clearance]. We tried to sell another system after flight 20 exploded, but certain government people resisted because they wanted the money to go to a different system.

So for several years we were at the forefront trying to sell an additional system and the real culprit was [name redacted] trying to stop it. He had political and other connections and was able to do it, believe it or not. There was intense rivalry between various agencies and military services for reconnaissance-related contracts.

Later on after our system was finished, we did crack some areas. When the new NRO was organized all of reconnaissance was put there, rather than have competition between the Navy, the Air Force, and the CIA for contracts and money. With the reorganization there were guys from all of these services or agencies all working together as one team. That ended the rivalry over reconnaissance.

TRANSFERRING THE PROGRAM

When the program was transferred from the CIA to the Air Force in the early 1970s, the transition for management and operation of Hexagon went smoothly. The Air Force wanted overall control since they already were running all of the launches. Dave Bergannini was the program manager for the Air Force and Tom O'Neill was the deputy—those guys were very good. Tom was located onsite in Danbury and was able to interface directly with all of the key managers and engineers there. The Aerospace Company, otherwise known as Circle A, was involved in the program from the early days and they made sure that the transition went well and that no technical stuff was dropped.

Developing and Presenting the S-Cubed

I came from a company that pioneered detectors, Fairchild Camera. Fairchild recognized that CCDs developed by Bell Labs would be unbelievable technology for the military. Fairchild bought that technology from Bell Labs and brought those people who were associated with that project to Fairchild on the West Coast. We on Long Island at Fairchild were working on the military applications.

When I came to Perkin-Elmer I was working with Rod Scott (Perkin-Elmer's chief scientist) and another guy working on how to determine attitude on our system. I came up with the idea of using CCDs. Larry Cassidy and several others worked on how to use solid-state detectors and what kind of accuracies we could achieve. Terry Galagher was also involved in that study. We showed that we could do the job. Colonel Dan Hutchinson at NRO was in charge of that program, and he helped get that through along with a guy named Bob Mahoney who was responsible for mapping and developing maps with an organization in St. Louis. With them and our own people we were able to sell the concept for S-cubed. It became a hell of a program.

One problem on S-cubed was that we had two doors, one for each camera located on the frame. The doors flipped open (like a toilet seat cover) for operation and were there so that if the S-cubed camera were to be accidentally aimed directly at the sun, it would prevent the sun from staring down the barrel of the camera and causing damage.

We were always worried about their opening and closing and whether they would stay shut, which would have been a disaster. I know we had some concept of bolts that could blow the doors off but we were worried about the danger of that. Anyway there were other cameras flying with the doors shut and they couldn't open them up. So the government was very leery of something with doors.

Anyway we launched the first S-cubed and the doors opened fine. We took some pictures and had a review on the West Coast in front of colonels and generals on the condition of the main camera and S-cubed. Lockheed gave a dissertation on how their equipment was working, and the downrange tracking people were there also to give their report. Colonel McChristian of the Air Force was in charge of the system. I remember him asking, "Abraham, we got the doors open. What are you going to do?" I said "We are going to leave them open and never close them again." "Okay, sit down," He replied.

We had done a study showing there wasn't any particular position that we could get into that the sun could go right down the barrel of the S-cubed and cause any problems. In case it did happen, there was no harm to the detector. So the doors remained open and that made the Air Force feel great.

Appendix F

Perkin-Elmer Personnel

Phil Pressel

Perkin-Elmer Corporation staff that worked on the Hexagon program over its years of existence, from study and proposal through its completion consisted of over 1200 people. There were many departments and sections and I have tried, with the help of some of my colleagues, to list as many as possible. In some cases I was given some very old organization charts. I also used some old Perkin-Elmer phone books, but many of the names came from my and colleagues' memories.

Wherever possible I have listed the names as I recall them working in certain departments. Some later transferred to other departments, and some are listed in general as I do not recall where they worked. My apologies go to those that I did not mention or whose names I misspelled.

Systems Engineering

Roger Arguello
A. Bardelli
Laura Benanti
Ron Betini
Henry Blazek
Dick Carricato
Larry Cassidy
Ann Chapleau
Mary Ann Chimera
Paul Convertito
Bob Crout
Bill Deane
Len Izzo
John Garrish
Frank Gerlack
Gene Gessner
Bob Gordon
Lee Gurski
Karl Herring
Dick Hess
Franz Holtzer
J Hudak
Nick Izzo
Peter Johansson
Len Kaplan
R Kaprinski
Mike Krim
Bob Kudola
Gerry McCue
Adolph Meier
Ken Miklus
Danny Milsom
Bill Moore
Dan Orlitzki
John Patterson
Karl Reseck
W. Rippel
George Shaboo
R. Sichel
Sam Sixby
Anthony Squillane
Roy Stoll
J. Virgulak
R. Warren
Bill Welch
Peter White

Copyright © 2013 by the American Institute of Aeronautics and Astronautics, Inc., All rights reserved.

Mechanical Engineering

Tom Alesi
Tony Bassaro
Bert Boyson
Paul Brickmeier
Ed Butterely, Jr.
Steve Cakouros
Karl Carter
Dick Chester
Don Cowles
Howard Cummings
Nick DeFillipis
Joe Dunn
Carol Farmer
Marge Ferris
Don Fish
Neil Freeman
Dean Fuller
Roger Gaulin
Sharon Gerback
Gerry Goldstein
Glen Grabow
Paul Jones
Charlie Karatzas
Fred Klein
Wayne Kohman
Bob McLaughlin
Len Molaskey
Bill Newell
Marvin Parnes
Phil Pressel
Dennnis Preston
Joe Prusak
Millie Roman
Walter Roman
Frank Sileo
Alex Siziew
Ken Stout
Joe Vollaro
Grahan Wallace
Connell Ward
Paul Yettito

Electrical Engineering

Gene Aha
Stan Anhouse
Roger Arguello
Ed Belles
Ken Bolin
George Bossers
Dick Carricato
Art Chickneas
Richard Fink
Walter Fowski
George Garafola
Al Gauthier
Al Gayhart
Gerry Gottesman
Warren Huber
Harvey Kaplan
Frank Krausz
Richard Labinger
Gerry LeBlanc
Ray Mittenthal
Arnie Negrin
Carol Nurmi
Lew Pennel
Bob Perliss
Bill Rappoport
Richard Rau
Joe Savarese
Frank Scinicariello
Stan Schwartz
John Shamaly
Chet Sokol
John Stein
Jay Sturdevant
Peter Van Raalte
Vic Varjassy
Jeneau Voller
bob Wada
Steve Whiting
Richard Whittlesey
Bob Williamson
Ed Wolpin
Al Yanik
Marty Yellin

Design & Drafting

Don Boughton
Ed Brelling
Fred Busch
Joe Cernak
Joe D'Agostino
Doc DeBattista
Frank Eno
Jim Foley
Ralph Forsberg
Ed Gaal
Al Gasper
Gil Gonzalez
George Grauel
Jack Gearin
Tom Hayden
John Keating
Steve McKinley
Don Mooney
Tom O'Mara
Len Petruzzelli
Sue Rieger
Dan Schiro
Ray Schneider
Bob Streeman
Fred Woodger
Joe Zelle

Manufacturing

John Anderson
Andy Babasick
Mike Barrachina
Andy Bentsen
George Boland
Frank Bonomo
Paul Bouffard
Al Bronico
John Butterkofer

Dick Cambrey
J. Canaperi
T. Christolini
Bob Churchill
Frank Citrin
John Connic
Gerry Cordone
Ron Corona
Pete Cronin
Dick Cummings
Richard Dick
Charlie Dipple
Ray Douglass
Roy Erickson
A. Fachini
J. Feledie
Dave Foster
Dave Fournier
Ron Garofolo
Gene Gerety
Paul Gildea
Wes Gillingham
Frank Grabowski
Hans Haberland
Al Hammond
Dave Hansen
Ted Hoff
Sol Hollar
Joe Howard
John Hroziencik
Augie Izzo
Frank Johnson
Glen Johnson
Jim Kalley
K. Kearney
Peter Keilich
A. Kinsey
Jocko Kodz
Nick Krauss
Frank LaBarbera
Larry LaRue
Al Loewenstine
Tom Lynch
Jim Macerhacker
Keith Managan
Dick Marks
Tom McClung
George Meloni
Kent Merserve
Oscar Mesko
Frank Mikaitis
Pete Moran
Tony Mordino
W. Morris
George Murray
Joe Nash
Ed Nelson
Ed Nicholas
Ed Ogden
Bernie O'Neill
Al Osanitsch
Phil Pane
Jack Parsons
John Piela
Al Pirone
Walt Pleva
Lois Reed
Ted Reinheimeer
Bob Remele
Al Reynolds
Earle Sangster
Tony Scarpone
Woody Seal
Bernie Shanahan
Jack Shanley
Mars Shaw
R. Sibilio
Ralph Sisk
Jim Smith
Bill Snyder
Steve Sprindis
Henry Tagatac
R. Toher
Dwight Travis
Paul Vacarre
Rich Vertula
Lester Waldeck
Stan Walker
Jack Walsh
Ray Weber
Frank Wendler
J. Wesson
Dan Westcott

Optics

Walt Augustyn
Raul Casas
Don Chaloux
Lloyd McCarthy
Juan Races
Bud Rigby
Al Slomba
Walt Wieland

Test

Tom Andrew
Jon Aspinwall
George Aube
Charlie Backman
Bob Bantz
Howie Burg
Jim Bleakney
R. Bock
Chuck Bradley
Charlie Bryant
Dick Buhrman
D. Burdett
Mike Cacioppo
John Campbell
Ray Canapini
Fred Cannaverde
Pete Cavanaugh
Charlie Chedester
Tom Chuvala
Art Cocchia
Dick Conrad
Bill Courville
Don Crosby
Jack Cunniff
Vince Daniello
J. Davis
George Davon
Moe DeFina
Tom DellaCorte
Joe DeMaiorebus
Bob Dillman
John Doyle
M Durkin

Dave Eddy
Gerry Elm
Dick Erdtmann
Tom Erdtmann
Scott Ferris
Bob Fink
Duane Flowers
Sandra Fluskey
Bill Foley
Dick Frank
Don Gabriel
Ed Gaffney
Johnny Gailes
Vinnie Gaudiosi
Jim Gereg
Fred Gillespie
M. Gilvert
V. Hall
Henry Heintz
Jordan Jack
John Jackson
John Kane
Mike Kayak
Charlie Keller
Jim Kelley
Frank Kochan
Bob LaPenna
George Lauri
Mark Lawson
Gordon Lester
Norm Lieberman
Jim Livingston
Dick Loewenadler
Gene Lotocki
John Mahalik
Keith Managan
R Mariano
Matt Matrullo
Ed Matthews
Louise Mesko
George Mikk
T Miller
Joe Miraglia
Steve montvila
Bill Morelli
Ed Moskal
Art Napolitano
Gerry Neidhart
Arnie Neuman
John Noll
Bernie O'Neill
Tim O'Neill
T. Ornowski
Sam Palasciano
Joe Pantelone
A Pellegrino
Ward Petrie
Clovis Prendegast
Pat Purdy
R Quadrini
Rich Rhyins
O. Rossi
Regina Rourke
T. Rowley
Jimmy Sangster
Sven Sathre
Jim Schneider
John Schoenwald
Bill Sennello
Sam Sixby
Paul Smith
John Smitrovich
Don Stein
Dick Strempski
John Swatt
John Tyra
R Voytek
Gene Waldenmair
Dan Westcott
Al Westfall
Ed Witcomb
Leidy Williams
Norm Woodbury
Charlie Yorio
John Zarka

Quality Assurance

Al Abele
Ed Augustauskas
Dick Backus
Cliff Beers
P. Behuniak
Dick Biggar
Joe Biglarderi
Dick Birdsall
Helen Blonski
Gerry Braddon
Dave Burch
Fred Cannaverde
Bill Carr
Dick Conway
Don costello
John Daly
Bernie Dorogusker
Tom Dubos
Frank Effrece
Wade Englund
Bernie Fisher
Glen Grabow
Bob Grover
Ken Grumann
G. Havas
Gene Hickman
Joe Howard
Dean Kessler
Frank Krasnikas
Marvin Kurmper
Bill Letteris
Dick Licarie
Rick Licaris
Jim Livingstone
Bill Ludford
Keith Managan
Dick Marks
Fred Marra
Lisa Meade
John Menser
Lou Mims
Joe Miraglia
Bill Morris
Ed Newton
Kaude nielsen
Jim Nugnes
L. Read
Marty Rosenberg
Terry Rowley
Hank Sauer
Bill Schnitzle
Dick Sisk
Jim Stewart
Bob Storms

Paul Tardiff
Bob Tita
Paul Toupin
Duane Turoczi
Dick Wolderauer
Herb Weisblatt
Kassh White
Walt Wieland
Bob Whitesell
G. Yurchishin
Andy Zelle

Reliability

Joe Basevich
Oscar Berendsohn
Bob Bialecki
Nick Bove
Joe Cantonese
Don Cota
Lou Cristo
John Daiga
Gerry Elm
Frank Finnerty
Ted Grabarz
Joe Hoehl
Stan Karachuk
Frank Katona
Andy Kieras
Vic Klimaytis
Bill Krohn
Bill Kulas
Frank Labarbera
Jim Luckasavage
Louie Manez
Jim Monte
Pete Moran
Norma Murphy
Bill Nardella
Marylin O'Hair
Dick Paltauf
Gerry Park
Ron Pivarnik
Steve Prince
Ted Schaket
Bob Schneider
Gerry Segal
Bob Smith
Myron Smith
Bill Straub
Jessee Taylor
Bill Von Allen
Warren Wilczewski
Ernie Williams
Jim Woods

Technical Documents

Ron Beck
Don Bossa
Mary Bresmon
Lorraine Coffey
Peter Green
Pete Grusczak
Betty Grusczak
Ry Haithwaite
Stan Karpacz
Bob Ogden
Ralph Panagrosso
John Restivo
Frank Tudino
Jean Walker

Upper Management & Tech. Staff

Vic Abraham
Roger Arguello
Dick Babish
Wally Benson
Hank Bielawa
Joe Borst
Bruce Boyce
Earle Brown
Val Buonaito
Jules Cohen
Jack Cunniff
Tom Evans
Paul Foreman
Dan Genovese
George Hefferon
George Henderson
Karl Henderson
Karl Herring
Robert Hering
Bob Jones
Karl Kelley
Heine Kessler
Jim Kiraly
Stan Kishner
Frank Kissh
Charley Lapinski
Mike Maguire
Bernie Malin
Walt McCammond
Mike Mazaika
Chester Nimitz, Jr.
Tom Norman
Ed Offenhartz
Paul Petty
John Rawlings
Bill Rayford
Harry Robertson
Milt Rosenau
Frank Scott
Rod Scott
Bob Sorenson
Don Stroberg
Eph Tobin
Arnie Wallace
Mike Weeks

Security

Lee Coleville
Frank DeLuca
Shelley Ferber
Tex Hernandez
Fred Malsheimer
Bob Markin
Hoyt Tuttle

Others

James Acquilina
Bob Alicky
Gerald Ahern
Kniffin Allen
Theresa Angell
Ray Antunes
Tom Arconti
Don Avery
Bob Bantz
Bob Beauchesne
Cliff Beers
Ed Benedict
Ted Berthold
Randy Bock
Ed Broderick
Carol Brooks
Dennis Brooks
Earle Brown
Beverly Buonaiuto
Val Buonaiuto
Frank Bulsom
Howie Burg
Al Burr
Ed Butterly, Sr.
Russ Byes
Lauren Cadell
John Calderbank
Ray Canapini
Rocco Cascio
Bob Cerda
George Chaber
Ed Chack
Dick Chester
Bob Clarke
Grady Cochrane
Bill Coleville
Ed Collins
Clark Coughlin
Bruce Daab
Judy Dale
Maurice Damast
Frank Davis
Ed Deveaux
Bob DeVitto
Jack Dietter
Dom Dellavalle
Joe DeMajorebus
John Dolnier
Bernie Dotogusker
Jim Dunn
Larry Emmons
Wade Englund
Al Erickson
Ron Esposito
Lew Fagan
Tom Fahey
Carol Farmer
Debbie Ferber
Mike Ferrara
Bernie Fisher
Gerri Fisher
Vicki Forest
Jim Gereg
D. Gillotti
Norm Goralnick
George Gray
Robert Griffin
Sandra Grossi
Bob Grosso
George Gwilliams
Charley Hall
Terry Hansell
Jack Hawkins
Doug Heacock
Dick Hess
B. Howell
John Hroziencik
James Hull
George Huse
Dave Imbrogno
Nick Izzo
Jim Kalley
John Kane
Joe Karcsmar
Stan Karpacz
John Kelley
Jim Kelley
Marion Kery
Arlene Kershnar
Dick Kertesz
Marlene Kertesz
Jim Kiraly
John Knauer
Bill Krohn
Marvin Krumper
Ron Kudola
Marge Lantry
Carol Lemoine
Dale Lizi
Steve Luty
Don Mahon
Bob Malin
Jack Malloy
Dick Malone
Fred Marchionna
Joe McGivern
Jim McGivern
Hugh McGovern
Barbary McKenney
Betty McLung
Pat McWhirt
John Menti
Ken Miklus
Tom Miller
Dale Miller
Jim Miller
Aaron Mills
J. Mills
Bill Mitchell
Hank Moeller
Richard Molaskey
Lou Montagnino
Steve Montvila
William Morris
Bud Mortenson
Jim Mullen
Art Napolitano
Joe Nash
Arnie Negrin
Vince Nichols
Bob Norman
Tom Norman
Joe O'Connor
Bob Ogden
Dave Olson
Bernie O'Neill
J. O'Neill
Tim O'Neill
Harold Osgood
Henry Osterweiss
Bob Peck

El Pelky
John Pettibone
Bob Pickett
John Piela
Ron Pivarnik
Sherman Poultney
Jules Racenet
Bill Raiford
John Rawlings
Al Renna
Sue Rieger
Anthony Rigogliosos
Cookie Ristau
Harry Robvertson
Barry Robinson
Erwin Rosenthal
Pete Rourke
Bob Rowley
Bob Roylance
Steve Rudof
Len Russo
John Russo
Glen Ryning
Gerri Schatzle
Vincent Schirf
John Schneider
Dan Scott
Harvey Sellner
Bill Sennello
Phil Siegel
Gene Sferrazza
John Sivinski
John Sliech
Joe Sliech
Eilleen Smith
Tony Squilante
Jim Stewart
Bob Storms
Dave Stroberg
Lucy Sutton
Sherry Testa
Bob Tita
Bill Tita
Ed Torian
Tom Tracey
Lou Tracy
Tom Trapasso
Ben Troidle
Dick Tyler
Jim Ullom
Al Vaghi
John Virgulak
Joe Von Dietsch
Mary Lou Voytik
Bill Wakeling
Jim Walker
Lou Wanamaker
Connell Ward
Jim Ward
Jay Warden
Gary Weller
Herb Weisblatt
Ed Whitcomb
Tim Wilcox
Warren Wilczewski
Al Wissinger
Gerry Zimmerman

West Coast Field Office

John Adams
Gerry Anthony
Don Armstrong
Robert Bair
Paul Barber
Alma Bates
Olivia Beahm
Carol Beck
Dick Berg
Tom Bolich
Denise Bowen
Calvin Bronner
Chuck Bruner
Carmine Buoniuto
Tom Canale
Sharon Canfield
Bill Courville
George Cummings
Jim Davis
Tom Day
Billy Deahm
Jim Dimas
Ken Dudschus
Lincoln Endelman
Joe Falcone
Jim Florang
Jerry Fogel
Bill Fullerton
Chris Garcia
Larry Green
Ed Greene
W.C. Hale
Marvis Hansen
Frank Harrigan
John Hawkins
Aurie Heath
Dino Hendrickchev
Mert Herbst
Bob Himmel
Ginger Hirose
Dan Hogan
Wilma Hooker
Gary Howe
Bernard Janowski
Jose Jasrez
Art Kaldem
John Kauer
Cathy Kidder
Ken King
Paul Knowlton
Jay Koch
Pete Krump
Joe Kulpa
John Lenihan
Jim Lewis
Frank Limarzi
Lyle Lindsay
Jim Lobig
Bill Lynch
Lyle Lindsay
Angelo Mallazzo
Jim Manis
George Manolis
Joan Martin
Tony Martino
John McNerney
Bob Meyer
Bill McClafferty
Dick Melberger
Bert Nelson
John Nordstrom
Danae Pennell

Maxine Odom
Roy Odom
Bob O'Leary
Ed O'Neil
Tom O'Neill
Walt Parkola
Clarence Parks
Dominic Pascale
Dan Pennell
Bill Quinn
Fred Reardon
Harold Reinhardt
Chuck Rementer
Tom Ricketts
John Sandell
Mike Scheer
Larry Schiemer
Roger Scholes
Ollie Shortall
Ford Smith
Payton Smith
Ron Smith
Tom Squires
Mike Stivaletti
Gerry Strebe
Grover Takamoto
HB Tanner
Glen Theobold
Al Thompson
John thomson
Neil Thompson
Paul Travis
Ted Urban
Patti Warshauer
Jim Walls
Jim Welborn
Don Williamson
George Woods
Ed Woyshner
Bob Zarba
John Zarka

INDEX

Note: Page numbers in *italics* indicate figures.

Supporting Materials

To download supplemental material files, please go to AIAA's electronic library, Aerospace Research Central (ARC), and navigate to the desired book's landing page for a link to access the materials: arc.aiaa.org.

A complete listing of titles in the Library of Flight series is available from ARC. Visit ARC frequently to stay abreast of product changes, corrections, special offers, and new publications.

AIAA is committed to devoting resources to the education of both practicing and future aerospace professionals. In 1996, the AIAA Foundation was founded. Its programs enhance scientific literacy and advance the arts and sciences of aerospace. For more information, please visit www.aiaafoundation.org.